JN424725

2nd Edition

Abstract Algebra

Sun Tae Soh

Department of Mathematics
Myong Ji University

KYOWOO

Copyright ⓒ 2019 by Kyowoo Co., Ltd. Printed in Korea All rights reserved.
No part of this publication may be reproduced or transmitted
in any from of by any means, electronic or mechanical, including
photocopy, recording, or any information storage and retrieval system,
without permission in writing from the publisher.

Kyowoo Co., Ltd.
8, Yangnyeongsi-ro, Dongdaemun-gu,
Seoul, Korea 130-861,

Phone: 82-2-925-2861
F a x: 82-2-925-2860
e-mail: kyowoo@kyowoo.co.kr
http://www.kyowoo.co.kr

Printed in the Republic of Korea.
ISBN 979-11-251-0281-6 (93410)

₩ 24,000

To My Beautiful Image.

Contents

Preface to the Second Edition

This book, *Abstract Algebra*, was first published in 2007 and then was improved, as time goes, in the following impressions, up to 2012. Since more time has passed after that, with the fixing of some still-remaining typos and errors of mine as much as I could, I added in this second edition a little bit of new materials to the latest impression of my book. Thus, this edition would be called the final version of my book, *Abstract Algebra*.

As a remark of this preface, I would like to mention briefly what I had foreseen in 1999, 20 years ago, what will be the main aspect of Mathematics in the 21st century, that is included as an Appendix A: An Introduction to Algebraic Analysis,[1] of my book, because most of the people around the world nowadays recognize the importance of Computation Related Human Activities(CRHA), after watching AlphaGo's victory in a Go-game competition in Seoul, in the Spring of 2016. Since the main aspect of CRHA consists of (1) Symbolic Computation, i.e., Algebraic Analysis, (2) deductive reasoning or inductive reasoning for intellect's understanding, and (3) decision making process in order to reach a meaningful conclusion after finitely many steps in reasoning process.

I think it is very important for us, human-beings, to understand what each of these three actually means, and it is fairly well described in the Appendix A: An Introduction to Algebraic Analysis, of my book, although I did not mention them explicitly in terms of technical terminologies, because, as it is well-known that the Greek Philosophy of those philosophers such as Socrates (470 - 399 BC), Plato (428/427 or 424/423 – 348/347 BC), and Aristotle (384 - 322 BC) is based on the method of Mathematical proof, still nowadays only well-taught in the courses of Pure

[1] http://ch.catholic.or.kr/pundang/4/soh/mathnet_973071.htm

Mathematics, which had already well been developed by the Pythagorean School of 6th century BC, CRHA are also definitely based on the method of Mathematical proof, still nowadays only well-taught in the courses of Pure Mathematics.

I believe that people well know that there is no contradiction among any facts, as long as they are all Mathematically proved to be true, and hence this is the main reason why the method of the Mathematical proof is very important to us, human-beings, as long as we reason why. Another thing is that people usually do not make a mistake in deriving a true conclusion through a deductive reasoning from sound hypotheses, but people sometimes derive a false conclusion through an inductive reasoning even from sound hypotheses that is NOT a mathematical inductive reasoning. One should also recognize that Artificial Intelligence is born to have such a capacity but it inevitably has, like us human-beings, a defect in its inductive reasoning process.

I think the Appendix A: An Introduction to Algebraic Analysis and the first chapter of my book is in fact well-suited for the beginners who want to learn how to better think and understand CRHA in the Computer Age, the 21st century, because their main underlying theme is not about any particular subjects in Mathematics, but rather about how we should think and reason to derive a meaningful conclusion in an axiomatic manner.

Toward this direction, some amount of my research works and papers on finding or establishing a new relationship among objects, i.e., concepts/ideas, in Computation Related Human Activities(CRHA), that are meaningful to us human-beings, existing generically or specially among the given ideas/concepts, under the environment of internet-based Artificial Intelligence, which have been produced by the author for a period of more than recent 15 years, are available to the readers through the internet access.[2]

Sun T. Soh
March 2019

[2]http://ch.catholic.or.kr/pundang/4/soh/a_list_of_research_outputs.htm

The rest of the preface is the original preface to the first edition of this book:

Preface [to the First Edition]

We recall that we extend our number system, the set $\mathbb{N} = \{1, 2, 3, ...\}$ of all natural numbers, to the set $\mathbb{Z}$ of all integers from our sincere efforts to obtain, for instance, a solution for X for a given linear equation, $X+n = 0$ for a given number $n \in \mathbb{N}$. Here, we note that the word, natural, means "God-given".

This is perhaps our original source for the right definition of -3 as "a (new, I mean, non-natural) number whose addition to the given number 3 results in the 0, i.e., emptiness", i.e. $(-3) + 3 = 0$. Of course, we "know" that we have only one such new number if ever it exists within our domain of thoughts. Thus, given a number 3, "*a linear relation of being equal to each other*", $(-3) + 3 = 0$, exactly tells us that we "know" either -3 from 0 or 0 from -3, as long as we live in $\mathbb{Z}$, i.e., we accept these (non-natural numbers) as new members of our numbers.

Similarly, we extend our number system, $\mathbb{Z}$, to the set $\mathbb{Q}$ of all rational numbers from our sincere efforts to obtain, for instance, a solution for X for a given linear equation, $nX = 1$ for a given non-zero number $n \in \mathbb{Z}$. Similarly, this process thus allows us to have the right definition of the new number $\frac{1}{3}$. It is a number whose multiplication to the given number 3 results in 1, i.e., the unit. Of course, we have only one such new number if ever it exists within our domain of thoughts. Thus, given a number 3, "*a linear relation of being equal to each other*", $\frac{1}{3} \cdot 3 = 1$, exactly tells us that we "know" either $\frac{1}{3}$ from 1 or 1 from $\frac{1}{3}$, as long as we live in $\mathbb{Q}$, i.e., we accept these (another non-natural numbers) as new members of our numbers. The reason why we mention these very basic stuffs is that it in fact allow us to understand what is Algebraic Analysis, or equivalently, Symbolic Computation.

For instance, 0.3333333333 is never the same as $\frac{1}{3}$ since

$$0.3333333333 \cdot 3 = 0.9999999999 \neq 1.$$

In Numerical Analysis, however, we simply approximate $\frac{1}{3}$ as 0.3333333333 and we regard 0.3333333333 as $\frac{1}{3}$ in all later computation, although it is mathematically absolutely wrong to treat 0.3333333333 in actual computation as if it were $\frac{1}{3}$.

Why did we accept this? Perhaps this has been the case because Numerical Analysis was studied extensively before enough development of software technologies especially for Symbolic Computation. Consequently, most students and professors in Engineering College still do not know or even understand the main differences between Numerical Analysis and Algebraic Analysis, i.e., Symbolic Computation. One of these two subjects does not replace the other. Rather, each compensates the other's weak points, so that they help each other.

What are we supposed to do when we are given a system of linear equations in several unknowns? We just need to learn how to solve such a linear system and this is all about Linear Algebra. Thus Linear Algebra will tell us what's really going on when we attempt to solve a system of linear equations. For instance, once you learn this subject a little bit, you can tell others explicitly and affirmatively whether such a system has a solution or not, and if so, you can actually tell others about how to compute all the solutions of the system in a systematic manner. Understanding this part of Linear Algebra is very important because almost every subject in Linear Algebra can be expressed as a system of linear equations.

Then what about a system of non-linear, i.e., polynomial equations in several unknowns? Our understanding is that virtually no books on either Linear Algebra or Algebra attempt to mention this, and consequently most students who take Linear Algebra or Algebra courses never have a fair chance of thinking about its usefulness and importance in advance.

In our treatment of this book, we try to study most of the first few materials of Modern Algebra, i.e., groups and multiplication-related, reverse-multiplication-related computation-oriented materials, for instance, polynomial rings in one unknown, as much as we could. Like other books for Modern Algebra, we begin our discussion axiomatically starting from groups, but as we go further, we try hard to move toward computational aspects of the subjects. The speed of our axiomatic treatment on the subject is rather slow but very solid and efficient, in the sense that quite many useful concrete examples are provided throughout the book. The word "Abstract" in the title of this book actually means "Axiomatic" and the crux of this book may be Chapter 4 and Chapter 6.

This book has grown out of my teachings on the subject for several years in my University. In writing this book, I took M. Artin's famous book, *Algebra*, as a basic model. On the other hand, this book is also prepared as an introduction to the book, *Algebraic Structures*, of my teacher, late Professor George R. Kempf (1944 - 2002) at JHU, with hidden but strong computation-oriented motivations. (In fact, I was the proof-reader of his book, *Algebraic Structures*.)

Here are author's comments on each chapter of the book.

In Chapter One, we introduce the most basic and hence common terminologies in Algebra with some motivational explanations. Since the materials are treated in an axiomatic manner, it is helpful for the beginners to have a chance of looking at the bare skeleton of the whole subject as roughly as possible, which is the main purpose of preparing our Chapter One.

Thus, Chapter 2 is the real beginning of this book. In this chapter, we introduce the very basic materials of Group Theory from nothing, i.e., axiomatically so to speak, thus the materials in this chapter are essential to later chapters of this book. The basic tendency of this chapter is purely Algebraic, so that discussing some concrete examples quite often is very important in order not to loose motivations and the right direction where we are heading for.

In Chapter 3, we study the action of a group on a set in a very systematic manner. Our treatment is basically geometry-oriented in the sense that the tools for analyzing the symmetries of certain well-known figures in the plane or the space and those objects invariant under such symmetries are well developed in a quite efficient manner, thus the title: Toward Abstract Symmetry. Thus, Chapter 3 is in some sense a dual to Chapter 2 in Geometry.

In Chapter 4, we apply everything what we have learned in the previous chapters in order to further discuss symmetries in the space, which reaches up to the proof of the three Theorems of Sylow for finite groups. In particular, we prove that A_5 is simple, using the symmetries of the icosahedral group of rotations of the regular dodecahedron in the space, following somewhat F. Klein's idea on the subject. The reader will find in this chapter that proofs of quite many results are very elegant. For example, the proofs of Sylow's Theorems are very conceptual and short. Although our treatment of the subject in this book is axiomatic, the reader will surely convince himself or herself that the concrete formula like the Class Equation for a group under the action of conjugation enriches the subject of axiomatic and hence abstract nature very much, and hence such a computation-oriented motivation does make axiomatic approach very meaningful to us.

Chapter 5 is the beginning the second half of this book. In Chapter 5, we study the very basic materials of the Ring Theory. We even construct the natural numbers, the integers, polynomial rings, and try to prove all the usual properties of numbers, from nothing, i.e., axiomatically. Thus, the materials covered in this chapter are essential in understanding all kinds of multiplication-related phenomena around us. We also discuss the concept of relations among elements in a ring, which eventually enables us to introduce new rings in a systematic manner, called Adjunction of Elements, as a silent starting point of Field Theory, which is, though, not treated in this book.

In Chapter 6, we discuss aspects of factorizations in some detail, thus the title: Factorizations. Since factorization process is, for instance, a reverse process of multiplication of numbers or polynomials, the subjects covered in this chapter are by no means easy ones, because no easy and efficient methods of factorizing out a product of numbers are yet discovered. In particular, the reader should note that Kronecker's ingenious opening theory of the Theory of Field Extensions is in fact heavily based on the Theory of Factorization, which is in turn based on the concept of Relations among Generators, covered in Chapter 5 of this book. Thus, if someone is interested in construction process of algebraic objects through computation or Mathematical Algorithms which are usually not mere protocols so to speak, it is very important to study factorization-related phenomena in a great detail. In particular, we discuss in some detail how PID's are different from UFD's, because ignorance of them in the past caused a serious confusion and hence trouble to quite many talented mathematicians. For example, the polynomial ring with more than one unknowns over a field is a UFD, but is never a PID, and this caused us human beings a terrible headache, for almost about 2500 years, in dividing out a given polynomial in more than one unknowns by several given ones in terms of quotients and remainders somewhat in a coherent manner in such a way that *the remainder is uniquely computed.*

As for the coverage of the materials, author's another book with Title: *Computational Algebra*, is the next one to study. There, those subjects in Algebra such as the Structure Theory for finitely generated abelian groups through the Theory of Modules, Field Theory, and Galois Theory are well developed from a very computation-oriented point of view, based on the Theory of Groebner Bases, which is in turn fully studied in author's another book with Title: *Algebraic Analysis I*, in which the readers will find, for instance, more computation-related advanced subjects in Invariant Theory and Computational Algebraic Geometry.

Unless we approximate numbers, which is very common in Numerical Analysis, the nature of all the objects under our mathematical consideration is exact. However, this usually requires us to multiply more than one unknowns and thus this requires us to study polynomial rings in several (i.e., more than one) variables over an (algebraically closed) field, called

"Computational Algebraic Geometry", which is beyond the scope of the main themes of this book. On the other hand, even Personal Computers nowadays already provide us with an excellent Hardware means for symbolic computation. For more about a balance between axiomatic approach and computational approach, one may refer to the Appendix of this book.

For more about Computer Algebra Systems, one may visit our homepage:

http://trinitas.mju.ac.kr/intro2softwares.html .

There, you will further find very useful information for users on CAS, Maple, Reduce, Matlab, etc.[3] Also,

http://trinitas.mju.ac.kr/intro2InetCompu.html

describes an important aspect of the availability of Internet environment to everybody in this wonderful world of 21st century.[4]

[3] http://ch.catholic.or.kr/pundang/archive/4/intro2softwares.html
[4] http://ch.catholic.or.kr/pundang/archive/4/intro2InetCompu.html

CHAPTER 1

A Few Fundamental Concepts in Algebra

1. Introduction

In this introductory chapter, we are going to review a few very basic concepts and definitions in Algebra. Before we start what constitutes Algebra, we would like to have a look at what had happened in our numbers, first. This will enable the readers to easily recognize that most of the defining axioms for terminologies currently used in Algebra are in fact taken from those ones satisfied by natural numbers, integers, or rational numbers. After that, we are going to discuss in a naive manner about a *metamathematics* of the process of proving a mathematical statement to be mathematically true. For this, we may need to recall the fact that one *can* drive a car even without knowing at all how a car operates mechanically, *as long as one manages to learn how to drive a car!*

2. Extensions From Natural Numbers to Complex Numbers

Every human being knows what are natural numbers, otherwise he is not one of us. That is, we know that if we add the number one to the number one, then we have the number two, and if we add the number one to the number two, then we have the number three, etc., i.e., we can keep going on this indefinite process to obtain another natural number repeatedly. This is our first incidence of recognizing an infinite set, the set of all natural numbers,

$$\mathbb{N} = \{1, 2, 3, 4, ...\},$$

and nobody knows exactly why and how we have recognized natural numbers in this way.

Moreover, we "know" that every non-empty subset S of $\mathbb{N}$ has a least element in it, i.e., we "accept" this as a fact without its proof. We also "know" that **the associative laws**

$$a+(b+c)=(a+b)+c \quad \text{and} \quad a\cdot(b\cdot c)=(a\cdot b)\cdot c.$$

hold for both addition and multiplication of any three natural numbers a, b, and c. We also "know" that **the distributive laws**

$$a\cdot(b+c)=a\cdot b+a\cdot c \quad \text{and} \quad (a+b)\cdot c=a\cdot c+b\cdot c$$

hold for natural numbers, etc. This means that when we say that we "know" natural numbers, we in fact means that we also accept these axioms as facts, and we are ready to apply these axioms in our later arguments on numbers. Since everybody knows about what constitutes natural numbers these numbers are some times called God-given numbers. That is, "natural" means "God-given", and conversely.

Then how about the negative numbers? Suppose a wife has prepared a tonight's dinner for 3 family members, including herself. But, her husband brought 2 friends with him to have dinner together at his house tonight. Then, his wife will quickly figure out that there will be no food for her, and this will make her unhappy. The situation is abstracted as follows. Given the number 3, she basically figures out the relationship that

$$(-3)+3=0,$$

where recognizing the zero, 0, or the negative number, -3, requires the recognition of the other. The point here is that she recognizes this relation first in her head; and then the usefulness of the negative number -3 and/or the number 0 (here meaning that nothing left to her) as the terminal of this relationship. Here, by definition,

DEFINITION 1.1. 0 *is a number whose addition to any number n is any number n itself.*

Since this kind of arguments is always possible for any other number n, our slow abstraction leads us to the introduction of the set of integers,

$$\mathbb{Z} = \{..., -3, -2, -1, 0, 1, 2, 3, ...\}.$$

Thus, *by definition*, the number -3 is a number whose addition to the natural number 3 results in the number 0 (i.e., nothing left). We then easily check that if there were somewhere such a number -3 then it must be unique. But since this number is not an element of $\mathbb{N}$, we need to extend the set $\mathbb{N}$ of natural numbers to the set $\mathbb{Z}$ of integers, so that we can do such an arithmetic.

This can be more abstractly expressed as follows: Solve the following equation for the unknown X:

$$X + 3 = 0.$$

Since its solution -3 is not one of the natural numbers, we had to extend our number system from $\mathbb{N}$ to $\mathbb{Z}$ in order to satisfy arithmetic demands of our everyday life like the above example.

With this information, we can prove that

LEMMA 1.1.

$$-(-1) = 1.$$

PROOF. For, $-(-1)$ exactly means that it is a number whose addition to the number -1 results in the number 0 (i.e., nothing left). That is,

$$-(-1) + (-1) = 0.$$

But, we also know

$$1 + (-1) = 0.$$

Then since there is only one number whose addition to -1 results in the number 0, we conclude that $-(-1) = 1$. □

Similarly, suppose we are given a 12-inch round pizza and 4 persons are going to equally share it. Then we obviously know how to divide it into 4 pieces of equal size and equal shape, and this establishes an abstract relationship,

$$\frac{1}{4} \cdot 4 = 1,$$

which equivalently says that the addition of the same $\frac{1}{4}$ six times is equal to the number 1, or equivalently it says that $\frac{1}{4}$ is a number whose multiplication to the natural number 4 is the number 1. Thus, by definition $\frac{1}{4}$ is a number whose multiplication to 4 results in the number 1 (i.e., a unit.) Here, **by definition**, 1 is a number whose multiplication to any number n is any number itself.

We then easily check that if such a number exists then it must be unique. Since this kind of arguments is always possible for any other number n, our slow abstraction leads us to the introduction of the set $\mathbb{Q}$ of rational numbers, which is an extension of $\mathbb{Z}$. This can be more abstractly expressed as follows: Solve the following equation for the unknown X:

$$X \cdot 4 = 1.$$

Since its solution $\frac{1}{4}$ is not one of the integers, we had to extend our number system from $\mathbb{Z}$ to $\mathbb{Q}$ in order to satisfy arithmetic division demands of our everyday life like the above example.

Moreover, we have

LEMMA 1.2.

$$0 \cdot 1 = 0.$$

PROOF. This is because, for any number n, we have

$$\begin{aligned} 0 \cdot 1 &= (n + (-n)) \cdot 1 \\ &= n \cdot 1 + (-n) \cdot 1 \\ &= n + (-n) \\ &= 0. \end{aligned}$$

□

On the other hand, Greek mathematicians thought that the set of positive rational numbers must be everything, i.e., there are no other numbers they can think of. But, this was a big mistake. For this, consider a rectangular triangle of two equal sides 1. Then even Babylonians in almost 4, 000 years ago knew (without proof) that there *is* a number whose square is equal to $1+1=2$, from the ruler and compass construction of magnitudes, i.e., positive numbers. This number is represented as the square root of 2:

$$\sqrt{2}.$$

LEMMA 1.3. *$\sqrt{2}$ is not a rational number.*

PROOF. Suppose that it is a rational number, say $\frac{a}{b}$ with $(a,b)=1$, i.e., a and b are relatively prime. Then that whose square is two exactly means that $a^2 = 2b^2$; hence a^2 is divisible by 2; hence a itself is divisible by 2; hence $4a'^2 = 2b^2$. This just implies that b is again divisible by 2, which is but a contradiction to our choice of a and b being relatively prime to each other. □

This can be more abstractly expressed as follows: Solve the following equation for the unknown X:

$$X^2 = 2.$$

Obviously, the disciples of Pythagoras should know about this proof, but history tells that they were forbidden officially to discuss this strange number, $\sqrt{2}$.

Another famous example is π. By its geometric definition, it is a ratio of two magnitudes, the length of circumference of a circle to that of the diameter of the same circle. Thus, if both of these magnitudes are rationals then π must be a rational number. Thus, the question is: when we choose the diameter of our circle to be the unit 1, the length (i.e., the magnitude) of the circumference of the same circle is whether rational or not.

Of course the right answer to this question took a pretty long time and we nowadays know that π is not a rational number. The history of mathematics tells us that pretty some time ago one mathematician spent his whole life to see if the decimal representation of π is periodic or not. (It is a fact that if its decimal representation is periodic, then it is a rational number.) Of course, he could not prove that it is periodic during his life time since by dividing infinitely many times for decimal fractions, he attempted to prove what he must prove its complete opposite, and another mathematician eventually proved that it is not a rational number at all, spending something like almost more than 20 years of his life time!

We note, unlike other examples as we have seen so far, that π is not accepted as a number as a solution of an equation containing an unknown X.

In fact, it took more than $1,800$ years since the Greek period for us to properly define the real numbers using the concept of limits, or limiting process, and it turns out that among real numbers, there are more irrational numbers (i.e., non-rational numbers) than rational numbers.

Having accepting the real numbers $\mathbb{R}$ as numbers, we still have a difficulty if we are asked to solve an equation for X:

$$X^2 + 1 = 0.$$

Obviously, there are no real solutions to this equation, because a square of any number is positive unless it is zero. On the other hand, if there is a number whose square is -1, then mathematicians in Italy of Renaissance period figured out that a very interesting thing happens.

3. A Few Remarks on Connectives

Before we go further, we would like to discuss how the logical connectives are used. In particular, we will discuss the right way of using implication in mathematics. Suppose we are given two input statements p and q. Then the truth value of $p \wedge q$, $p \vee q$ and $p \Rightarrow q$ as outputs are defined as follows:

p	q	$p \wedge q$	$p \vee q$	$p \Rightarrow q$
T	T	T	T	T
T	F	F	T	F
F	T	F	T	T
F	F	F	F	T

Thus, we know immediately the truth value of the outputs $p \wedge q$ and/or $p \vee q$, once we know whether each of the two inputs p and q are true or false.

But the situation is quite different when the usage of the $p \Rightarrow q$ is involved. Although it is defined as is in the above table, its usage in mathematics is the following:

p	$p \Rightarrow q$	q
T	T	T
T	F	F
F	T	T
F	T	F

Here, the only input is p and we want to verify (or prove) whether q is true or false as an output, through a mathematical proof, which corresponds to the step $p \Rightarrow q$. When our input (or hypothesis) p is true, since the truth value of both $p \Rightarrow q$ and q are exactly the same by definition, our final result (or output) q must be true (false respectively), as long as our argument, $p \Rightarrow q$, is true (false respectively).

p	$p \Rightarrow q$	q
T	T	T
T	F	F

On the other hand, when our input (or hypothesis) p is false, our output q can be either true or false regardless of our argument $p \Rightarrow q$ being true. For instance, any mathematical conclusion from a false hypothesis, like 3 divides 5, could be either true or false regardless of our argument

$p \Rightarrow q$ being true.

p	$p \Rightarrow q$	q
F	T	T
F	T	F

This is what it really says: Suppose you want to know whether a statement q is true or false. If you start from a false statement p and, via a completely error-free $p \Rightarrow q$ step, arrive at q, then, even though your argument or reasoning step $p \Rightarrow q$ is completely error-free, the very statement q is still true or false, which was but the initial state of the statement q!; hence you still do not know whether the statement q is true or false!

Now, here is our explanation on the so-called method of contradiction in our proof business and it goes as follows: Suppose we are able to reach a conclusion q through a true argument, i.e., a valid $p \Rightarrow q$ step, from a hypothesis p whose truth value is not yet verified, but on the other hand everybody is able to see that "that" conclusion q is in fact false, then the **only** conclusion we can draw is that the hypothesis p is false. This is in fact the exact meaning of the following table, where the truth value of p is the output.

p	$p \Rightarrow q$	q
F	T	F

EXAMPLE 1.1. *A good example of a proof by the method of contradiction is the proof of Lemma 1.3 above.*

4. A Few Remarks on Mathematical Statements

In mathematics, certain terminologies are not defined. For example, a “point” and a “line” are two such examples. But, we do understand what it means when one says either a line is a collection of points or the intersection of two lines gives us a point. Such terminologies are called **undefined terminologies**. They are terminologies for certain objects whose existence is undeniable to every human being, so to speak.

On the other hand, we can not verify the validity of certain statements. For example, “two parallel lines in the plane do not meet at all”, or "any two lines in the plane do meet at a point". Obviously, these two statements are contradictory to each other, but unfortunately we can not either prove or disprove independently that either one of them is true or false. A statement whose mathematical validity is for granted, i.e., we "believe" that it is mathematically true, is called an **axiom**. Thus, we have here non-comparable two axioms.

We then create a statement combining these undefined terminologies and axioms, using the **connectives**. Then a statement is called "**mathematical**" only when we are able to demonstrate whether it is mathematically true or not, but not both (in other words, half-true and/or half-false depending on different situations) under any situation at all. Thus, a mathematical statement should not have a double-meaning or a circumstantially inequivalent meaning under any situation at all.

The process of verifying (that is, establishing or justifying) a statement to be mathematically true is called a **proof**.

A statement which is proved to be true is called a **theorem**. Most of the time, theorems are called a **lemma**, a **theorem**, a **corollary**, or a **proposition**, or even a **scholium**, for the purpose of indicating and/or emphasizing its relation to other theorems.

5. A Few Remarks on Proofs

REMARK 1.1. *A few remarks on proofs are in order:*

(1) *If some concept (or a notion) has just been defined and we are asked to prove something concerning the concept (or the notion), we must use the definition of the concept (or the notion) as an integral part of the proof.*
(2) *The statement of a theorem consists of two parts: the* hypotheses *and the* conclusion. *If all the hypotheses are needed to prove the theorem, then, each hypothesis must be cited somewhere in the proof.*

(3) *If one happens to figure out an example where a statement is not true, then the statement is not a theorem. Such an example is usually called a* ***counterexample****.*

(4) *One should not tacitly assume that any hypothesis that is not explicitly stated, and one should not take for granted any quantifying words or phrases such as* only, for all, for each, *or* for some *that do not actually appear.*

(5) *A theorem of the form*

If hypotheses *then* conclusion

can not be proved by giving a special example where the hypotheses and conclusion are all true. For, one must show that for all *examples where the hypotheses are true, the conclusion is also true.*

(6) *Suppose that one wants to prove that*

 (a) *an element satisfying a certain property P exists and moreover*

 (b) *its mathematical existence is* unique.

 For this, one first needs to prove that there exists such an element, and then to prove its unique existence, one first assume that there are two elements r, s satisfying the same property P and then try to demonstrate that $r = s$.

(7) *A theorem of the form*

If P *then either* Q *or* R

is proved by showing that Q *(or* R, *respectively)) holds under the additional assumption that* R *(or* Q, *respectively) is not true.*

6. What is Algebra?

The subject Algebra has a long history in Mathematics. For instance, even in Greek period, people knew how to add, subtract, multiply, and divide the **magnitudes**, even if they did not know how to write down numbers efficiently and even if they did not have symbols for these four basic operations, $+, -, \times$, and $\div$. They did all their mathematical thinking in a verbal way, based on the so-called method of "Dialogues". Even with the Roman way of representing numbers people had a great difficulty in multiplying numbers. As time goes, people very slowly began to introduce several efficient symbols to represent numbers and the basic four operations on numbers, etc.

When the followings are what we are about to deal with, we call it **Algebra**.

- We treat both **numbers** such as $1, 2, 3, ...$, and **symbols** such as $a, b, c, \pi, x, y, z, ...$
- Some of the symbols, for instance, $a, b, c, ...$ etc., are already **known** to us and others, for instance, $x, y, z, ...$etc., are yet **unknown** to us, but they are treated as if they were numbers, regardless of being known or unknown to us.
- These numbers and symbols are connected using the four connectives: $+, -, \cdot$, and $\div$ (**Addition, Subtraction, Multiplication, and Division**, respectively).
- We "**equate**" two such expressions with $=$ (Equality) sign to have an equation.
- We apply the following rules (or principles) to an equation:
  - (**Addition rule**) Whenever we add a number or a symbol on one side of an equality, we have to do the same thing on the other side of the equality. [Whenever we subtract a number or a symbol from one side of an equality, we have to do the same thing on the other side of the equality.]
  - (**Multiplication rule**) Whenever we multiply [divide] a non-zero number or a symbol on one side of an equality, we have to do the same thing on the other side of the equality.
  - (**Substitution rule**) Whenever we substitute a symbol in one side of an equality with another symbol or number, we have to the same thing on the same symbol on the other side of the equality.

- We try to express any unknown symbol to be equal to a **polynomial expression** of the rest of known and unknown symbols and numbers.

REMARK 1.2. *We remark that the* ***decimal notation*** *of numbers such as* 12345 *and those* ***connecting symbols*** *such as* $+, -, \cdot, \div$, *and* $=$ *were not generally used almost until the end of* 15th *century. Before this time, most mathematical expressions are given as plain sentences, as are in Euclid's Elements. For a long time, the method of the so called "****Dialogues****" has been the best method of learning mathematics.*

7. Why (Abstract) Algebra?

The main objects of our study are polynomials in either one variable or more than one variables over a field k. Only the most elementary mathematical skills are required to manipulate polynomials, and it is really necessary to work within the larger frame of abstract, i.e. axiomatic, algebra in general.

The concept of abstract algebra arises from the observation that certain operations such as addition and multiplication can be performed on a variety of objects, such as numbers, polynomials, functions, or matrices, to name a few. Certain properties of these operations are often shared by different objects on which these operations are performed.

For example, we consider addition, multiplication, and division with remainder of integers which we will discuss very soon. We will also see that these operations can be performed not only with integers, but with polynomials in one variable over the field of rational numbers or that of real numbers (in fact, over any field), and that the same basis properties such as commutativity or the associativity hold.

Thus, if two sets share a structure and a property of that structure, then they also share all consequences of this property, and hence we would like to derive these consequences in a generalized setting. For example, all the results on integers we will discuss will have precise counter parts for polynomials in one variable, and this is exactly what abstract (i.e., axiomatic) algebra is all about.

8. What is Linear Algebra?

It is certainly a branch of Algebra with only one restriction: We **never** multiply unknown symbols or quantities together (which usually results in more unknown symbols). This means that linear algebra is about how to solve a system of linear equations, and it is about how to use such a computational method in handling other parts of mathematics as applications. Whenever this restriction is applied, it belongs to the subject, called **Linear Algebra**. Since computation itself has no taste at all, we usually need to understand what's really going on behind such computations. This means that we need to learn new concepts which enable us to study certain important properties and/or results in Linear Algebra.

9. What is Non-Linear Algebra?

On the other hand, the existence of a subject called Linear Algebra then also means the existence of a non-linear Algebra in general, where several unknown symbols or quantities are allowed to be multiplied together. This means that the business of doing Linear Algebra is conceptually included in that of doing Non-Linear Algebra.

For example, if X is unknown, then X^2 is another unknown that is more difficult to know than X itself, and X^3 is another unknown that is more difficult to know than X^2, and so on. The degree of difficulty may be seen graphically by drawing, for instance, the graphs of $y = x$, $y = x^2$, $y = x^3$, etc.

But the situation was changed dramatically at around the middle of the 20th century. Starting 1965 we are now able to solve a system of polynomial equations, thanks to Austrian Mathematician **Bruno Buchberger**'s discovery of an algorithm for computing the so-called "Groebner Basis" for a polynomial ideal in several variables over a field k.

In fact, his discovery of an algorithm for computing a Groebner basis is very important: It is a natural generalization of the algorithmic process of obtaining the row-reduced echelon form of a system of linear equations in Linear Algebra, and it is also somewhat a natural generalization of either finding the greatest common divisor of two polynomials in one variable or that of two natural numbers. Moreover, his theory is in fact a generalization of Gauss' famous result, known as "**the fundamental theorem of Algebra**":

THEOREM 1.1. *Every polynomial equation of degree d in one variable over the field $\mathbb{C}$ of complex numbers has exactly the degree d number of complex roots.*

This may need at least some further explanations.

When we are given a system of finite number of polynomial equations in one unknown, i.e., variable, over a field k, we do not need to attempt to solve common roots of these equations. Instead, we first compute their greatest common divisor $d(X)$ and then solve the single equation $d(X) = 0$ to figure out all the possible common roots of them, where the existence of solutions of $d(X) = 0$ is completely guaranteed by the Gauss' theorem above, and for different reasons we sometimes need to apply the famous numerical method, Newton's method, to figure out the solutions of $d(X) = 0$. In a later chapter, we will have a chance to review these elementary facts.

On the other hand, when we are given a system of polynomial equations in more than one unknown over even an algebraically closed field, finding their common solutions is generally very difficult, because the division process involved is very complicated, and here is the place where Buchberger's theory discovered in the year 1965 properly puts its stake. His theory is a natural generalization of the Gauss' theorem above, because it always **algorithmically** allows us to transform the given system of polynomial equations into another system of polynomial equations, of course, having the same solution set, in such a way that if there is no solution, then we can recognize it computationally, or else the last equation of the latter is a polynomial equation in one unknown, and the rest of the equations of the latter system is of the form which allows us to apply the so-called backward substitution process, so that we can now computationally figure out all the solutions of the system of polynomial equations over an algebraically closed field k. It took about $2,500$ years to discover such an important algorithm by Bruno Buchberger since the Greek figured out the proof of finding the greatest common divisor of two natural numbers, which is fully described in Euclid's *Elements.*

One of the main difficulties in solving a system of polynomial equations in more than one unknown is that the polynomial ring $k[X, Y]$ in two unknowns X and Y over a field k is a Unique Factorization Domain (*UFD*) but it is not a Principal Ideal Domain (*PID*) in general.

EXAMPLE 1.2. *We will later prove that $k[X, Y]$ is a UFD. But, the maximal ideal (X, Y) in $k[X, Y]$ is never generated by a single polynomial in $k[X, Y]$, because there is no non-constant polynomial in $k[X, Y]$ which divides both polynomials X and Y, irreducible over k. Thus, $k[X, Y]$ is not a PID. [Remark: We will introduce all the needed definitions and terminologies very soon.]*

Because of this, we can not uniquely determine quotients and a remainder of division of a polynomial $f(X, Y) \in \mathbb{Q}[X, Y]$ by non-zero polynomials $g_1(X, Y)$ and $g_2(X, Y)$:

$$f(X, Y) = a_1(X, Y)g_1(X, Y) + a_2(X, Y)g_2(X, Y) + r(X, Y).$$

For example, exactly because of this kind of difficulties, it is not easy to decide whether a given polynomial $f(X, Y)$ is a member of an ideal I that is generated by two polynomials $g_1(X, Y)$ and $g_2(X, Y)$. By definition, we know that

$$f(X, Y) \in (g_1(X, Y), g_2(X, Y)) \text{ if and only if } r(X, Y) \equiv 0.$$

Therefore, in order to computationally answer this fundamental belonging question we need to develop a computable means, that is, an algorithm, which can uniquely tell us whether the remainder $r(X, Y)$ in the above expression is the zero polynomial or not, in general. This problem is nowadays called **an ideal membership question**.

On the other hand, even a high school student knows how to divide out a polynomial $f(X) \in \mathbb{Q}[X]$ by a non-zero polynomial $g(X)$:

$$f(X) = a(X)g(X) + r(X)$$

in a unique manner. Thus, everybody knows whether a given polynomial $f(X)$ is a member of the ideal J generated by $g(X)$ or not by simply computing the remainder $r(X)$ using the division algorithm for polynomials in one variable.

In my another book, *Computational Algebra*, which follows this book, we will apply The theory of Groebner basis to compute the discriminant of cubic or quartic equations, for instance.

10. Several Basic Definitions in Algebra

Before we go into an algorithmic aspects of Non-Linear Algebra, let's have a look at several basic definitions in Algebra. In the long run, this will make the readers very comfortable in learning Algebra.

DEFINITION 1.2. *A* ***field*** *F is an integral domain where every non-zero element a is invertible, i.e., for each $a \neq 0$ there exists b in F such that $ab = 1$.*

Such an a is called a **left inverse** of b and such a b is called a **right inverse** of a.

EXAMPLE 1.3. *$\mathbb{Q}$ = the set of all rational numbers, $\mathbb{R}$ = the set of all real numbers, and $\mathbb{C}$ = the set of all complex numbers.*

REMARK 1.3. *In fact, the Greek knew $\mathbb{R}$ = the set of all real numbers, first, without defining it rigorously. They thought that every real number is represented as a ratio of two natural numbers, i.e., a rational number, but their attempt was wrong (See Lemma 1.3 above).*

DEFINITION 1.3. *An* ***integral domain*** *D is a commutative ring where the* ***cancellation law*** *holds, i.e., if $ab = ac$ and $a \neq 0$ then $b = c$ in D.*

REMARK 1.4. *Thus, an integral domain is such a place where you can cancel out a non-zero element, even without multiplying (mostly non-existing in D) its inverse on both sides of the equality under consideration. More precisely, an integral domain D is not a field in general but we can still "cancel out" the same non-zero element out of both sides of a given equality as if its inverse were multiplied on both sides of the equality, although the existence of its multiplicative inverse in D is not assumed at all. It is very interesting that the Greek knew about this important property of natural numbers .*

EXAMPLE 1.4. *Any fields, $\mathbb{Z}$ = the set of all integers, and $\mathbb{Q}[X]$ = the set of all polynomials in one unknown X with rational coefficients.*

DEFINITION 1.4. *A* ***commutative ring*** *K is a ring whose multiplication is commutative.*

EXAMPLE 1.5. *Any integral domains and the quotient ring $\mathbb{Z}/n\mathbb{Z}$ where n is a positive integer. Later, we will learn how to construct the quotient ring $\mathbb{Z}/n\mathbb{Z}$ in terms of Euclid's division algorithm.*

DEFINITION 1.5. *A* ***ring*** *R is an abelian group together with multiplication $R \times R \to R$ satisfying the following rules:*

(1) *(Existence of multiplicative identity) There exists 1 in R such that $r.1 = r = 1.r$ for any element r of R.*
(2) *(Associativity) $a(bc) = (ab)c$.*
(3) *(Distributive law) Multiplication is distributive over addition in both directions, i.e., $a(b + c) = ab + ac$ and $(a + b)c = ac + bc$.*

EXAMPLE 1.6. *The set of all $n \times n$ matrices, where n is one of the numbers,$1, 2, 3, ...$*

REMARK 1.5. *Thus, there are two binary operations in a ring and they are usually called "addition (denoted by $+$)" and "multiplication (denoted by $\cdot$)".*

DEFINITION 1.6. *An* ***abelian group*** *A is a group whose binary operation, $+$, is commutative, i.e., $a+b=b+a$.*

REMARK 1.6. *Its "additive identity" is usually denoted by* 0.

DEFINITION 1.7. *A* ***group*** *G is a set together with a binary operation $*$ satisfying the following rules:*

(1) *(Existence of identity) There exists an element e in G such that $g*e=g=e*g$ for each element g of G.*
(2) *(Associativity) $a*(b*c)=(a*b)*c$.*
(3) *(Invertibility) Every element of G is invertible.*

EXAMPLE 1.7. $\mathbb{Z}$.

REMARK 1.7. *We "**know**" a **set** S only when we know each of its elements. When have a finite set, we have no problem at all. But, most of the sets we are going to deal with in mathematics are infinite sets. For example,*

$$\mathbb{N} = \text{the set of all natural numbers } 1, 2, 3, \ldots$$

*is the most important infinite set whose infiniteness we all know. We recall that the traditional meaning of English word "natural" is "God-given". Natural numbers are also called "**counting** numbers" or "**ordering** numbers" for an obvious reason that they are always used in counting and/or ordering objects.*

REMARK 1.8. *If you read the above definitions backward, you will be able to read them in a proper logical order in defining them.*

Thus, there could be no binary operation on a set, we have one binary operation on a group G, or we have two binary operations $+$ and $\cdot$ on a ring R.

We need to define additional ones:

DEFINITION 1.8. *A* ***vector space*** *V over a field F is a module over the commutative ring F.*

EXAMPLE 1.8. $\mathbb{R}^2$ *= the real plane, and* $\mathbb{R}^3$ *= the Euclidean 3-space.*

DEFINITION 1.9. *A (left)* ***module*** *M over a commutative ring K is an Abelian group together with an operation (called scalar multiplication): $. : K \times M \to M$ satisfying the following rules: for each a, b in K and for each m, n in M,*

(1) $1.m = m$
(2) $(ab).m = a.(b.m)$
(3) $(a + b).m = a.m + b.m$
(4) $a.(m + n) = a.m + a.n$

EXAMPLE 1.9. *Any commutative ring K is a module over itself.*

EXAMPLE 1.10. *Any abelian group A is a module over $\mathbb{Z}$.*

DEFINITION 1.10. *An element a of a commutative ring R is a* ***zero divisor*** *if $a \neq 0$ and there exists $b \neq 0$ such that their product $ab = 0$.*

LEMMA 1.4. *A commutative ring R is an integral domain if and only if R has no zero divisors.*

PROOF. Suppose that R is an integral domain, that is, it satisfies the cancellation law. If $ab = 0$ and $a \neq 0$, then since

$$ab = 0 = a \cdot 0$$

holds, we cancel out a from both sides to conclude that $b = 0$,Thus, there are no non-zero b satisfying $ab = 0$ where $a \neq 0$. Conversely, if it has no zero divisors, it exactly means that $ab = 0$ implies either $a = 0$ or $b = 0$. Suppose that $ab = ac$ and $a \neq 0$. Then we have

$$a \cdot (b - c) = 0.$$

Then since $a \neq 0$ by hypothesis, $b - c$ must be zero, i.e., $b = c$, which is all about the cancellation law. □

DEFINITION 1.11. *A* ***binary operation*** *on a set S is a map,*

$$S \times S \longrightarrow S,$$

from their Cartesian product $S \times S$ to S. A binary operation is sometimes called ***a law of composition****.*

We use maps to compare sets:

THEOREM 1.2. *Let* $f\colon X \longrightarrow Y$ *be a map.*

(1) f *is one-to-one, i.e.,* ***injective****, if and only if it has a left inverse.*
(2) f *is onto, i.e.,* ***surjective****, if and only if it has a right inverse.*

PROOF. Suppose that f is injective. For any y in the image of f we have only one preimage x, so we define this y to be sent back to the x. For any y in Y outside the image of f, we define this y to be sent to the same particular element x_0 of X of our choice. This defines a map $g\colon Y \longrightarrow X$ such that their composition $g \circ f$ is the identity map: $X \longrightarrow X$. This proves the $\Rightarrow$ direction of the first assertion. The converse of the first assertion is obvious.

Now suppose that f is surjective. Then we note that the domain X of the function f is partitioned into the disjoint union of various non-empty preimages of the elements $y \in Y$. Thus, by the Axiom of the choice function, which is equivalent to the Axiom of choice, we can choose an element out of each preimage of elements of Y under the map f, and this defines a function $g\colon Y \longrightarrow X$ such that $f \circ g$ is the identity map: $Y \longrightarrow Y$. This proves the $\Rightarrow$ direction of the second assertion. The converse of the second assertion is obvious. □

REMARK 1.9. *In particular, we note that our proof also tells us that such inverses are not uniquely determined.*

DEFINITION 1.12. *A map is called* ***bijective*** *if it is both injective and surjective.*

COROLLARY 1.1. *A map* f *is bijective if and only if it has both a right inverse and a left inverse. When this is the case, any right inverse of* f *is equal to any left-inverse of* f. □

COROLLARY 1.2. *A map* f *is bijective if and only if it has a (two-sided) inverse* f^{-1}. *When this is the case, any two inverses of* f *are the same, and*

$$(f^{-1})^{-1} = f. \qquad \square$$

DEFINITION 1.13. *A map ϕ from a module M to a module N over the same commutative ring K is called a* ***linear*** *map or linear transformation of modules if for each $m, m' \in M$ and for each $\alpha \in K$,*

$$\begin{aligned} \phi(m+m') &= \phi(m)+\phi(m'), \\ \phi(\alpha m) &= \alpha\phi(m) \end{aligned}$$

hold.

DEFINITION 1.14. *An* ***isomorphism*** *of two modules M and N is a linear map from M to N that is also one-to-one and onto.*

DEFINITION 1.15. *A map ϕ from a commutative ring R to another commutative ring S is called a* ***homomorphism*** *of rings if for each r and $r' \in R$,*

$$\begin{aligned} \phi(r+r') &= \phi(r)+\phi(r'), \\ \phi(rr') &= \phi(r)\phi(r'), \text{ and} \\ \phi(1) &= 1 \end{aligned}$$

hold.

DEFINITION 1.16. *An* ***isomorphism*** *of two commutative rings R and S is a homomorphism from R to S that is one-to-one and onto.*

The readers are strongly recommended to memorize the above definitions in the given order.

There is one theorem which is really essential to us.

THEOREM 1.3. *(**Mathematical Induction**) A set of natural numbers that contains the number 1 and the number $n+1$ whenever it contains a number n must be the set of all numbers.* □

DEFINITION 1.17. *A* ***subfield*** *F of a field k is a subring of k that is also a field.*

DEFINITION 1.18. *A* ***subring*** *S of a commutative ring R is a subset of R satisfying*

(1) $1 \in S$, *where* 1 *is the (only) multiplicative identity element of* R.
(2) $a - b \in S$, *whenever* $a, b \in S$.
(3) $ab \in S$, *whenever* $a, b \in S$.

DEFINITION 1.19. *A* ***subgroup*** H *of a group* G *is a subset of* G *satisfying*

(1) $e \in H$, *where* e *is the (only) identity element of* G.
(2) $ab \in H$, *whenever* $a, b \in H$.
(3) $a^{-1} \in H$, *whenever* $a \in H$.

DEFINITION 1.20. *An (left)* ***ideal*** I *of a commutative ring* R *is an Abelian subgroup of* R *under addition satisfying*

$$ra \in I,$$

whenever $a \in I$ *and* $r \in R$.

In other words, a subset I of a commutative ring R is an ideal if it satisfies the following conditions:

(1) $0 \in I$.
(2) If $a, b \in I$ then $a + b \in I$.
(3) If $a \in I$ and $r \in R$, then $ra \in I$.

11. A Concluding Remark

We may recall that the Well-Ordering Principle, which is equivalent to the Axiom of Choice in Mathematics or the so-called Zorn's lemma (which turned out to be falsely called as a lemma), is in fact a simple generalization to any set X of a property P satisfied by the set $\mathbb{N}$ of all natural numbers.

The Well-Ordering Principle says that every set X has an ordering $>$ on it such that it satisfies the following property P: each non-empty subset S of X has a least element in S itself.

But, the main problem with this Principle is that the set (and consequently, any non-empty subset) of all natural numbers $\mathbb{N}$ is, to our best knowledge until now, the only known set equipped with an ordering $>$ on it satisfying such a property P, so that a "logical" generalization of such a very particular property of $\mathbb{N}$ to an arbitrary non-empty set X is quite difficult for us to reasonably accept it, although we nowadays accept it as an Axiom in Mathematics. For instance,

EXAMPLE 1.11. *Even on the set $\mathbb{Z}$ of all integers, we have so far failed to figure out such an order $>$.*

But starting in the first quarter of the 20th century, mathematicians (mostly working in Real Analysis) first began to recognize the necessity of the Zorn's lemma in finishing the proof of some fundamental theorems, and then at around the middle of the of the 20th century it has been proved that this falsely called axiom, the Zorn's lemma, is in fact independent of any other axioms in mathematics and moreover, that the three are in fact all equivalent to each other.

The application of this axiom is really needed in Algebra. For instance,

EXAMPLE 1.12. *We still do not know how to prove that every vector space over a field has a basis without using the Axiom of Choice.*

EXAMPLE 1.13. *Another such an example where the application of the Zorn's lemma is definitely needed is the proof that each ideal in a commutative ring is contained in a maximal ideal.*

The main problem with any one of these equivalent axioms in reality is the following: once it is used for establishment of an existence of a certain mathematical object, the method applied, nonetheless, will never ever tell us how to construct such an object at all, and consequently, we really must know when any one of them is actually applied in the proving process in order for us not to be mistakenly involved with an impossible attempt of constructing the claimed mathematical object. In other words, we believe that the axiom itself and all of its proved consequences are mathematically true, but this acknowledgement is nothing to do with the construction process of a certain mathematical object of which existence is guaranteed by such a logical, namely, non-computational, axiom.

Exercises for Chapter 1

EXERCISE 1.1. *Prove that to say that the cancellation law holds in a commutative ring R is the same as to say that R has no zero-divisors, i.e., If $ab = 0$ for $a, b \in R$ then one of either a or b is zero.*

EXERCISE 1.2. *Can you define a field consisting of only 2 elements?*

EXERCISE 1.3. *Can you define a field consisting of only 3 elements?*

EXERCISE 1.4. *Can you define a field consisting of only 4 elements?*

EXERCISE 1.5. *Can you see that every Abelian group is a $\mathbb{Z}$-module?*

EXERCISE 1.6. *Derive a formula for the sum:*

$$S_5(n) = 1^5 + 2^5 + \dots + n^5,$$

and prove that your formula holds for each $n \in \mathbb{N}$. (Hint: Consider the difference $(k+1)^6 - k^6$ when k runs through from 1 to n.)

EXERCISE 1.7. *Can you explain a few of the main differences between exact computation and numerical computation?*

EXERCISE 1.8. *When and why do you use software packages like MAPLE, REDUCE, which are nowadays called Computer Algebra System, or Matlab?*

EXERCISE 1.9. *For numerical computations, a few methods of parallel computation had been developed and standardized: OpenMP and MPI. Are there any standardized methods for parallel computation for exact computations?*

CHAPTER 2

Groups

1. The Definition of a Group

In Chapter One, we have officially defined a group in an axiomatic manner:

DEFINITION 2.1. *A **group** G is a set together with a binary operation, or a law of composition,*

$$* : G \times G \to G$$

satisfying the following rules:

(1) *(Existence of identity) There exists an element e in G such that $g * e = g = e * g$ for each element g of G.*
(2) *(Associativity) $a * (b * c) = (a * b) * c$.*
(3) *(Invertibility) Every element of G is invertible.*

REMARK 2.1. *History tells us that Galois (1811 - 1832) was the first mathematician who recognized the importance of **symmetry** on the set of mathematical objects, and it was he who provided concrete examples (i.e., Galois groups in Galois Theory for a finite field extension) and consequently this urged others to the study on the group theory, especially on a classification of finite (abelian) groups, which was initially launched by Kronecker (1823 - 1891), a student of Eduard Kummer (1810 - 1893). It is known that Dedekind (1831 − 1916), the last Ph.D. student of Gauss (1777 - 1855), coined the official definition of a group given as above.*

EXAMPLE 2.1. *Here are some examples of groups:*

(1) *the set $\mathbb{Z}^+$ of all integers under addition,*
(2) *the set $\mathbb{R}^\times$ of all non-zero real numbers under multiplication,$\times$*

(3) *the set* $\mathbb{R}^+$ *of all real numbers under addition, and*
(4) *the set* $GL_n(\mathbb{R})$ *of all invertible matrices of size* $n \times n$ *with real entries under matrix multiplication.*

In this section, we are going to provide a more detailed explanation on the terminologies used in the definition of a group.

DEFINITION 2.2. *By* **a law of composition** *on a set* S*, we mean a rule for combining pairs* a, b *of elements of* S *to get another element, say* p*, of* S *in a unique manner, namely, it is a map*

$$f : S \times S \to S$$

sending any pair $(a, b) \in S \times S$ *to an element* $p \in S$*.*

The original models for this notion are addition and multiplication of real numbers. As long as notation is concerned, the functional notation $p = f(a, b)$ is not very convenient for laws of composition. Instead, the element obtained by applying the law to a pair (a, b) is usually denoted using a notation resembling those used for multiplication or addition:

$$p = ab, \ a \times b, \ a \circ b, \ a + b, \text{ and so on,}$$

a choice being made for the particular law in question. We usually call the element p the **product** or **sum** of a and b, depending on the notation chosen.

EXAMPLE 2.2. *Here is the first example of a law of composition, the matrix multiplication on the set* S *of matrices of the same size.*

We will use the product notation ab most frequently, and anything we prove with product notation can be rewritten using another notation, such as addition. It will continue to be valid, because the rewriting is just a change of notation.

It is important to note that the symbol ab is a notation for a certain element of S, but it is generally very difficult to figure out the original elements a and b explicitly out of their product ab.

EXAMPLE 2.3. *Let* $a = \begin{bmatrix} 1 & 2 \\ 0 & 3 \end{bmatrix}$, $b = \begin{bmatrix} 1 & 0 \\ 2 & 1 \end{bmatrix}$. *Then their product* ab *is* $\begin{bmatrix} 5 & 2 \\ 6 & 3 \end{bmatrix} = c$, *but it is very difficult to recover simultaneously both the exact* a *and* b *from the result* $ab = \begin{bmatrix} 5 & 2 \\ 6 & 3 \end{bmatrix}$.

On the other hand, it is very easy to uniquely recover b *from* $ax = c$, *because the matrix* a *is invertible:* $b = a^{-1}c$.

In general, the reverse process of multiplication is usually called a ***factorization*** *process, and it is one of the most difficult and consequently important subject of study in Algebra. For more about it and its consequences, refer to our next book, titled with "Computational Algebra".*

NOTATION 2.1. *When discussing groups in general, we will use multiplication notation. It is customary to reserve additive notation* $a+b$ *for commutative laws of composition, namely, when* $a+b = b+a$ *for each* $a, b \in S$.

REMARK 2.2. *Multiplicative notation carries no implication either way concerning commutativity of the law of composition, unless otherwise mentioned explicitly.*

DEFINITION 2.3. *A law of composition is called* ***associative*** *if the rule*

$$(ab)c = a(bc) \qquad \text{(associative law)}$$

holds for each $a, b, c \in S$, *and* ***commutative*** *if the rule*

$$ab = ba \qquad \text{(commutative law)}$$

holds for each $a, b \in S$.

EXAMPLE 2.4. *One can easily see that matrix multiplication is associative but not commutative.*

In additive notation the associative law is

$$(a+b)+c = a+(b+c),$$

and in functional notation it is

$$f(f(a,b),c) = f(a, f(b,c)).$$

This ugly formula illustrates the fact that functional notation is not convenient for algebraic manipulation.

On the other hand, the associative law is more fundamental than the commutative law, as the following examples explains:

EXAMPLE 2.5. *Let T be a set and f, g be functions (i.e., maps) from T to itself. As usual, we denote $g \circ f$ the composed map $x \longmapsto g(f(x))$. Then the rule*

$$(g, f) \longmapsto g \circ f$$

is a law of composition on the set $S = Maps(T,T)$, the set of all maps $T \longrightarrow T$. One then easily checks that "this" law of composition is associative but is not commutative.

Now, we want to define the product of a string of n elements of a set S:

$$a_1 a_2 \cdots a_n = ?$$

There are various ways to do this using the given law of composition, which tells us how to multiply two of them. For example, when $n = 4$, we have in total 5 different ways to combine the same elements a_1, a_2, a_3, a_4:

$$((a_1a_2)a_3)a_4,\ (a_1(a_2a_3))a_4,\ (a_1a_2)(a_3a_4),\ a_1(a_2(a_3a_4)),\ a_1((a_2a_3)a_4).$$

It can be proved by induction that if the law is *associative*, then all such finite products are the same, and this allows us to speak of *the* product of an arbitrary string of finite number of elements of S.

PROPOSITION 2.1. *Suppose that an* associative *law of composition is given on a set S. Then there exists a unique way to define, for every integer n, a product $[a_1 \cdots a_n]$ of n elements $a_1, ..., a_n$ of S satisfying the following three properties:*

(1) *the product $[a_1]$ of one element is the element itself;*

(2) *the product* $[a_1a_2]$ *of two elements is given by the law of composition,* $[a_1a_2] := [a_1][a_2]$ $(= a_1a_2)$;

(3) *for any number* i *such that* $1 < i < n$, *we have*

$$[a_1 \cdots a_n] := [a_1 \cdots a_i][a_{i+1} \cdots a_n].$$

PROOF. (Induction on n). The product in question is defined by (1) and (2) for $n \leq 2$, and it does satisfy (3) when $n = 2$. By induction hypothesis we know how to define the product of r elements for any $r \leq n-1$, and this product is the unique product satisfying (3). For $r = n$, we define the product of n elements by the rule

$$[a_1 \cdots a_n] := [a_1 \cdots a_{n-1}][a_n],$$

where the terms on the right side are those already defined by induction hypothesis. Then the formula we have just defined satisfies (3) in particular when $i = n-1$. If $i < n-1$, then we have

$$\begin{aligned}
[a_1 \cdots a_n] &= [a_1 \cdots a_{n-1}][a_n] && \text{(by our definition)} \\
&= ([a_1 \cdots a_i][a_{i+1} \cdots a_{n-1}])[a_n] && \text{(by induction hypothesis)} \\
&= [a_1 \cdots a_i]([a_{i+1} \cdots a_{n-1}][a_n]) && \text{(by associativity of the law)} \\
&= [a_1 \cdots a_i][a_{i+1} \cdots a_n] && \text{(by induction hypothesis).}
\end{aligned}$$

Thus, the formula we defined indeed satisfies (3).

Now, if a product satisfying these three properties exists, the formula we have defined gives the product again by induction on n because it is (3) when $i = n-1$; hence the product is unique, if exists, and this completes the proof. □

NOTATION 2.2. *From now on, we will drop the brackets and denote* the unique *product by*

$$a_1 \cdots a_n.$$

DEFINITION 2.4. *An* ***identity element*** *for a law of composition on a set* S *is an element* e *of* S *such that*

$$ae = a \qquad \textit{and} \qquad ea = a, \qquad \textit{for each } a \in S.$$

LEMMA 2.1. *There could be at most one identity element.*

PROOF. If e, e' were two such elements, then

$$e = ee' = e'.$$

□

NOTATION 2.3. *Often the identity element is denoted by* 1 *if the law of composition is written multiplicatively, or by* 0 *if it is written additively.*

EXAMPLE 2.6. *Matrix multiplication and composition of functions both have an identity. For* $n \times n$ *matrices it is the identity matrix* I *and for* $Maps(T, T)$ *it is the identity map.*

The elements do not need to be related to the numbers 1 and 0, but they share the property of being identity elements for their laws of composition.

Suppose that our law of composition on S has an identity, and let us use the symbol 1 for it.

DEFINITION 2.5. *An element* $a \in S$ *is* ***invertible*** *if there exists another element* $b \in S$ *such that*

$$ab = 1 \qquad \textit{and} \qquad ba = 1.$$

LEMMA 2.2. *The inverse of an element is unique, if it exists.*

PROOF. If b, b' are two such elements, then

$$b = 1b = (b'a)b = b'(ab) = b'1 = b'.$$

□

NOTATION 2.4. *The unique inverse of* a*, if it exists, is denoted by*

$$a^{-1}.$$

LEMMA 2.3. *The product of two invertible elements* a, b *is again invertible, and*

$$(ab)^{-1} = b^{-1}a^{-1}.$$

PROOF. Since we have

$$(b^{-1}a^{-1})(ab) = b^{-1}(a^{-1}a)b = b^{-1}(1b) = b^{-1}b = 1,$$

and similarly $(ab)(b^{-1}a^{-1}) = 1$, by the uniqueness of the inverse of ab, we finish the proof. □

NOTATION 2.5. ***Power notation*** *may be used for an associative law of composition:*

$$\begin{aligned} a^n &= \underbrace{a\cdots a}_{n \text{ times}} \qquad (n \geq 1), \\ a^0 &= 1 \qquad \text{provided the identity exists} \\ a^{-n} &= a^{-1}\cdots a^{-1} \qquad \text{provided } a \text{ is invertible.} \end{aligned}$$

LEMMA 2.4. *The usual rules for multiplication of powers hold:*

$$a^{r+s} = a^r a^s \qquad \text{and} \qquad (a^r)^s = a^{rs}.$$

PROOF. Exercise. □

REMARK 2.3. *It is not advisable to introduce fraction notation*

$$\frac{b}{a}$$

unless the law of composition is commutative. This is because it is not clear from the notation whether the fraction stands for ba^{-1} *or* $a^{-1}b$*, and these two elements may be different.*

NOTATION 2.6. *When additive notation is used for the law of composition, the inverse of* a *is denoted by* $-a$*, and the power notation* a^n *is replaced by the notation*

$$na := \underbrace{a + \cdots + a}_{n \text{ times}}$$

as with addition of real numbers. Similarly,

$$(-n)a := \underbrace{(-a) + \cdots + (-a)}_{n \text{ times}}.$$

Thus $(-n)a = (-a) + \cdots + (-a) = n(-a)$ holds.

It is customary to denote the group and the set of its elements by the same symbol.

DEFINITION 2.6. *An* ***abelian group*** *is a group whose law of composition is commutative.*

NOTATION 2.7. *Additive notation is used for abelian groups.*

Here are a few easy examples of abelian groups:

EXAMPLE 2.7.

(1) $\mathbf{Z}^+$*: the integers, with addition;*
(2) $\mathbb{R}^+$*: the real numbers, with addition;*
(3) $\mathbb{R}^\times$*: the non-zero real numbers, with multiplication;*
(4) $\mathbb{C}^+$*: the complex numbers, with addition;*
(5) $\mathbb{C}^\times$*: the non-zero complex numbers, with multiplication.*

PROPOSITION 2.2. *(Cancellation law) Let a, b, c be elements of a group G. If $ab = ac$, then $b = c$. If $ba = ca$, then $b = c$.*

PROOF. For, every element of a group is invertible. □

If an element a is not invertible, the cancellation law need not hold:

EXAMPLE 2.8. $0 \cdot 1 = 0 \cdot 2$ *but* $1 \neq 2$.

$$\begin{bmatrix} 1 & 0 \\ 0 & 0 \end{bmatrix}\begin{bmatrix} 1 & 0 \\ 2 & 1 \end{bmatrix} = \begin{bmatrix} 1 & 0 \\ 0 & 0 \end{bmatrix}\begin{bmatrix} 1 & 0 \\ 0 & 3 \end{bmatrix},$$

but

$$\begin{bmatrix} 1 & 0 \\ 2 & 1 \end{bmatrix} \neq \begin{bmatrix} 1 & 0 \\ 0 & 3 \end{bmatrix}.$$

Exercises for Chapter 2.1

EXERCISE 2.1. *Prove that a group has exactly one element x such that $x^2 = x$.*

EXERCISE 2.2. *Prove that if each element x of a group G satisfies $x^2 = e$, then G is an abelian group.*

EXERCISE 2.3. *Prove that the n-th roots of unity, i.e., the roots of $x^n = 1$ in the unit circle in the complex plane is a group.*

EXERCISE 2.4. *Let G be a finite group. Prove that some power of each element of G is e.*

EXERCISE 2.5. *Prove that if a group G has an even number of elements then there exists an element whose square is e.*

2. Two non-trivial Examples: GL_n and S_n

NOTATION 2.8. *Let*

$$GL_n(\mathbb{R}) \qquad \textit{and} \qquad GL_n(\mathbb{C})$$

denote the set of all $n \times n$ invertible matrices with entries in $\mathbb{R}$ and $\mathbb{C}$, respectively. We sometimes denote any of them by

$$GL_n = \{n \times n \textit{ matrices } A \textit{ with } \det A \neq 0\}.$$

One can easily check that it is a group, called the $n \times n$ **general linear group**.

NOTATION 2.9. *Let*

$$S_n$$

denote the set of all permutations, that is, invertible functions, on a set $\{1, 2, ..., n\}$.

One can easily check that it is a group, called the **symmetric group**. Because there are $n!$ number of permutations of a set of n elements, S_n has $n!$ elements.

S_2 consists of two elements, the identity permutation e and the transposition τ which sends 1 to 2 and 2 to 1, and hence $\tau^2 = e$. Thus, the group law, composition of functions, is described by the fact that e is the identity element and by the *relation*

$$\tau\tau = \tau^2 = e.$$

S_3 consists of 6 elements, and to describe this group we pick two particular permutations. Let x denote the cyclic permutation, i.e., the one sending

$$\begin{aligned} 1 &\mapsto 2, \\ 2 &\mapsto 3, \\ 3 &\mapsto 1, \end{aligned}$$

and y denote the transposition, i.e., the one sending

$$\begin{aligned} 1 &\mapsto 2, \\ 2 &\mapsto 1, \\ 3 &\mapsto 3. \end{aligned}$$

Thus, x and y satisfy the rules $x^3 = 1$ and $y^2 = 1$, respectively, and moreover x and y are *related* by the rule:

$$x^2y = yx.$$

We then also note that the three rules

$$x^3 = 1, y^2 = 1, x^2y = yx, \tag{2.1}$$

are sufficient for computation in the group S_3, because any product of the elements x, y and of their inverses can be brought to the form x^iy^j with $0 \leq i \leq 2$ and$0 \leq j \leq 1$. For instance,

$$x^{-1}y^3x^2y = x^2yx^2y = x^2(yx)xy = x^2(x^2y)xy = x(yx)y = xx^2yy = 1.$$

Thus, we can write a complete **multiplication table** for S_3 as follows:

(2.2)

	1	x	x^2	y	xy	yx
1	1	x	x^2	y	xy	yx
x	x	x^2	1	xy	yx	y
x^2	x^2	1	x	yx	y	xy
y	y	yx	xy	1	x^2	x
xy	xy	y	yx	x	1	x^2
yx	yx	xy	y	x^2	x	1

In particular, the table says:

COROLLARY 2.1. *S_3 consists of two elements, x, x^2, of order 3, three elements, y, xy, yx, of order 2.* □

Because of this, the rules are called **defining relations** for the group S_3.

LEMMA 2.5.

$$\begin{aligned} S_3 &= \{1, x, x^2, y, xy, x^2y\} \\ &= \{x^iy^j : 0 \le i < 3, 0 \le j < 2, x^3 = 1, y^2 = 1, yx = x^2y\}, \end{aligned}$$

where 1 denotes the identity permutation.

PROOF. Just compute the products. □

From these observations, we also see:

COROLLARY 2.2. *S_3 is not an abelian group.*

PROOF. This is because $yx = x^2y \neq xy$. □

REMARK 2.4. *In fact, S_3 is the* smallest *non-abelian group. This requires more explanation: if G is a group of prime order p, then it is cyclic and hence automatically abelian, which is a corollary of Lagrange's Theorem, Corollary 2.8: The order of a subgroup of a finite group G divides the order of G. A group of order 4 is called the **Klein 4 group** V, since it has three subgroups of order 2 and is generated by the relations*

$$x^2 = 1, y^2 = 1, xy = yx,$$

so it is automatically abelian.

One reason that the general linear group GL_n and the symmetric group S_n are so important is that many other groups are contained in them as subgroups.

REMARK 2.5. *Toward this direction, one may refer to, for instance, Definition 3.14 (a permutation representation of a group G) and Definition 3.16 (a matrix representation of a group G).*

Exercises for Chapter 2.2

EXERCISE 2.6. *Prove that $GL_n(\mathbb{R})$ is a group.*

EXERCISE 2.7. *Prove that S_n is a group.*

EXERCISE 2.8. *Give an example of 2×2 matrices A, B such that $A^{-1}B \neq BA^{-1}$.*

EXERCISE 2.9. *Make a multiplication table for the Klein 4 group V and verify that it has three elements of order 2.*

EXERCISE 2.10. *Determine the number of elements of order 2 in S_4.*

EXERCISE 2.11. *Determine the order of every element in the dihedral group D_4.*

EXERCISE 2.12. *In the dihedral group D_5 of regular pentagon (Definition 3.6), prove that every element except the identity has order 2 or 5.*

EXERCISE 2.13. *(*) Prove that D_6 is isomorphic to $S_3 \times S_2$.*

EXERCISE 2.14. *Show that the relations*

$$x^4 = 1, y^2 = x^2, y^{-1}xy = x^{-1}$$

define a group of order 8, called the "quaternion group" (Example 2.15).

3. Subgroups

DEFINITION 2.7. *A subset H of a group G is called a* ***subgroup*** *of G if it satisfies the following conditions:*

(1) *(Closure) $ab \in H$ for each $a, b \in H$.*
(2) *(Identity) $1 \in H$.*
(3) *(Inverse) $a^{-1} \in H$ for each $a \in H$.*

REMARK 2.6. *The first condition (1) says that the law of composition on the group G is again used as the law of composition on H, called the* ***induced law of composition.*** *The second and third conditions (2) and*

(3) say that H is a group with respect to this induced law. Here, we do not need to mention associativity because it carries over automatically from G to H.

We note that every group has two obvious subgroups: the whole group and the subgroup $\{1\}$, consisting of the identity element alone. A subgroup is said to be a **proper subgroup** if it is not one of these two.

EXAMPLE 2.9.

(1) *The set T of invertible upper triangular 2×2 matrices*

$$\begin{bmatrix} a & b \\ 0 & d \end{bmatrix} \qquad (a \neq 0, d \neq 0)$$

is a subgroup of $GL_2(\mathbb{R})$.

(2) *The set of complex numbers of absolute value 1, namely, the set of points on the unit circle in the complex plane, is a subgroup of $\mathbb{C}^\times$.*

LEMMA 2.6. *If H are K are subgroups of a group G, then so is their intersection $H \cap K$.*

PROOF. Just apply Definition 2.7. □

An easy exercise of this lemma is:

COROLLARY 2.3. *$H \cap K$ is a subgroup of H and of K.* □

It is also clear that $H \cap K$ is the largest subgroup of G contained in both H and K.

The following proposition determines all the subgroups of the additive group $\mathbb{Z}^+$ of integers.

PROPOSITION 2.3. *For any integer $b \in \mathbb{Z}$, the subset of $\mathbb{Z}$,*

$$b\mathbb{Z} = \{n \in \mathbb{Z} : n = bk \text{ for some } k \in Z\},$$

is a subgroup of $\mathbb{Z}^+$. Moreover, every subgroup H of $\mathbb{Z}^+$ is of the type $H = b\mathbb{Z}$ for some integer $b \in \mathbb{Z}$.

PROOF. It is easy to verify that $b\mathbb{Z}$ is a subgroup of $\mathbb{Z}^+$, so we leave its proof as an exercise.

Now, let H be a subgroup of $\mathbb{Z}^+$. Thus, it contains the identity element 0 of $\mathbb{Z}^+$. If $H = \{0\}$, then $H = 0\mathbb{Z}$, so the case is settled. Otherwise, there exists a non-zero element a in H. If a is negative then $-a$ is positive and is also in H since H is a group. Let a be the smallest positive number in H. Then obviously, we have

$$a\mathbb{Z} \subseteq H.$$

In what follows, we are going to prove the reverse inclusion: $H \subseteq a\mathbb{Z}$; whence proving that $H = a\mathbb{Z}$, as asserted.

For this, we apply division with remainder to write any $n \in H$ in terms of a and the unique remainder r:

$$n = aq + r,$$

with $0 \leq r < a$. Since both n and a are elements of H, their difference $r = n - aq$, which is strictly smaller than a, is also an element of H. Thus, by our choice of a, r must be zero; consequently $n = aq \in aH$, as required. □

Thus, we have lots of many different subgroups of $\mathbb{Z}^+$ by changing b in $b\mathbb{Z}$. Let a, b be two non-zero distinct integers and consider two subgroups $a\mathbb{Z}$ and $b\mathbb{Z}$. We are going to show that there are two special subgroups of $\mathbb{Z}^+$: the one that is the largest subgroup among those subgroups of $\mathbb{Z}^+$ contained in both $a\mathbb{Z}$ and $b\mathbb{Z}$, and the another that is the smallest subgroup among those subgroups of $\mathbb{Z}^+$ containing both $a\mathbb{Z}$ and $b\mathbb{Z}$.

LEMMA 2.7. *The intersection $a\mathbb{Z} \cap b\mathbb{Z}$ of $a\mathbb{Z}$ and $b\mathbb{Z}$ is a subgroup of $\mathbb{Z}^+$ which is the largest one among those subgroups of $\mathbb{Z}^+$ contained in both $a\mathbb{Z}$ and $b\mathbb{Z}$.*

PROOF. It is very routine to verify it, so we leave it as an exercise. □

PROPOSITION 2.4. $a\mathbb{Z} \cap b\mathbb{Z} = \mathrm{lcm}(a, b)\mathbb{Z}$.

PROOF. We have: $c \in a\mathbb{Z} \cap b\mathbb{Z}$ if and only if $a|c$ and $b|c$ if and only if $\text{lcm}(a, b)|c$ if and only if $c \in \text{lcm}(a, b)\mathbb{Z}$. □

NOTATION 2.10. *Let a, b be two non-zero distinct integers. We define a set $a\mathbb{Z}+b\mathbb{Z}$ by*

$$a\mathbb{Z}+b\mathbb{Z} := \{n \in \mathbb{Z} : n = ar + bs \text{ for some } r, s \in \mathbb{Z}\}.$$

LEMMA 2.8. *The set $a\mathbb{Z}+b\mathbb{Z}$ is a subgroup of $\mathbb{Z}^+$ that is the smallest one among those subgroups of $\mathbb{Z}^+$ containing both $a\mathbb{Z}$ and $b\mathbb{Z}$.*

PROOF. Once we show the first part, then the second part is self-clear. Moreover, the proof of the first part is also very routine, so we leave it as an exercise. □

The subgroup $a\mathbb{Z}+b\mathbb{Z}$ is called the subgroup ***generated by*** a and b.

Then by Proposition 2.3, we know that this subgroup $a\mathbb{Z}+b\mathbb{Z} = d\mathbb{Z}$ for some $d \in \mathbb{Z}$. Now, the next immediate question is how to figure out this d:

PROPOSITION 2.5. *Suppose that*

$$a\mathbb{Z}+b\mathbb{Z} = d\mathbb{Z}.$$

*Then $d = \gcd(a, b)$, **the greatest common divisor** of a and b.*

PROOF. Since both a and b are in the subgroup $d\mathbb{Z}$, d divides both a and b. Now suppose that an integer e divides both a and b. Then both a and b are in $e\mathbb{Z}$, and consequently any integer $ar + bs$ in $a\mathbb{Z}+b\mathbb{Z}$ is also in $e\mathbb{Z}$. Thus, $d\mathbb{Z} \subseteq e\mathbb{Z}$. That is, d is divisible by e. This proves that d is the *greatest* common divisor of a and b. □

COROLLARY 2.4. $\gcd(a, b) = ar + bs$ *for some integers r and s.*

PROOF. This result is immediate from the equality: $a\mathbb{Z}+b\mathbb{Z} = d\mathbb{Z}$. □

REMARK 2.7. *We note that the above treatments do not tell us how we compute* $\gcd(a, b)$, *r and s. To compute* $\gcd(a, b), r$, *and s, we apply the very well-known **division algorithm of Euclid** of linear time efficiency,* *which was known to us since about BC* 300*. Here is the algorithm of Euclid for* $\gcd(a,b)$.

ALGORITHM 1. *(Euclid's algorithm for* $\gcd(a,b)$*) We may assume that* $b > a \neq 0$*. If* b *is a multiple of* $a \neq 0$*, then* $a = \gcd(a,b)$*. Otherwise, we continuously apply the above theorem to obtain:*

$$\begin{aligned} b &= q_1 a + r_1, \text{ with } r_1 > 0, \\ a &= q_2 a_1 + r_2, \text{ with } a_1 = r_1 \text{ and } r_2 > 0, \\ a_1 &= q_3 a_2 + r_3, \text{ with } a_2 = r_2 \text{ and } r_3 > 0, \\ &\quad \ldots\ldots\ldots \\ a_{k-2} &= q_k a_{k-1} + r_k, \text{ with } a_{k-1} = r_{k-1} \text{ and } r_k > 0, \\ a_{k-1} &= q_{k+1} a_k + 0, \text{ with } a_k = r_k > 0 \text{ but } r_{k+1} = 0. \end{aligned}$$

Then $r_1 > r_2 > \ldots > r_k > 0$ *and hence* $\gcd(a,b) = .\gcd(r_1,r_2) = \gcd(r_2,r_3) = \ldots = \gcd(r_k,0) = r_k$. □

This algorithm does work since

$$\gcd(b,a) = \gcd(b-aq,a) = \gcd(r,a) = \gcd(a,r),$$

and by rewriting these equations backwards starting from $\gcd(a,b)$, we can find those integers r and s in $\gcd(a,b) = ar + bs$.

REMARK 2.8. *We note that these* r *and* s *are not uniquely determined.*

EXAMPLE 2.10. *We know by inspection that* $\gcd(6,29) = 1$*. On the other hand,*

$$\begin{aligned} 29 &= 4 \cdot 6 + 5, \\ 6 &= 1 \cdot 5 + 1, \\ 5 &= 5 \cdot 1 + 0. \end{aligned}$$

Thus, $1 = 6 - 1 \cdot 5 = 6 - 1 \cdot (29 - 4 \cdot 6) = 5 \cdot 6 + (-1) \cdot 29$*; hence* $r = 5$ *and* $s = -1$*. Obviously, these* r *and* s *are not determined uniquely:* $1 = 5 \cdot 6 + \{(-29) \cdot 6 + 29 \cdot 6\} + (-1) \cdot 29 = (-24) \cdot 6 + 28 \cdot 29$*, where* $6 \cdot 29 = \operatorname{lcm}(6,29)$*, etc.*

LEMMA 2.9. *Let p be a prime number and let a, b be numbers. If p divides their product ab, then p divides at least one of them.*

PROOF. Suppose that the prime p does not divide a. Then $\gcd(p, a) = 1$; hence by the above corollary we have

$$1 = pr + as$$

for some $r, s \in \mathbb{Z}$. Multiplying both sides by b we see that p does divide b. □

DEFINITION 2.8. *Let G be a group. A subgroup H*

$$\begin{aligned} H \quad : \quad &= \{x^n : n \in \mathbb{Z}\} \\ &= \{..., x^{-2}, x^{-1}, x^0 = 1, x, x^2, ...\} \end{aligned}$$

of G generated by a single element x of G is called a **cyclic subgroup** *of G.*

H is clearly a subgroup of G, the smallest subgroup of G containing x. Since x^n is a notation for a certain element of G, it may happen that there are repetitions in the list. For instance, if $x = 1$, then all the powers $x^n = 1$, so that $H = \{1\}$. Otherwise, we have two possibilities: Either the powers of x are all distinct elements, or not.

In the first case, the subgroup H is called **infinite cyclic**.

Otherwise, as n is increased from 1 some powers x^n is the same as the smaller power x^m with $n > m$. Then, $x^{n-m} = 1$ by Cancellation law (Proposition 2.2), and hence there exists a non-zero power of x which is equal to 1.

LEMMA 2.10. *Let S be the set of those integers n such that $x^n = 1$. Then*

(1) *S is a subgroup of $\mathbb{Z}^+$.*
(2) *$S = m\mathbb{Z}$, where m is the smallest positive integer in S.*

PROOF. Part (1). If $m, n \in S$, then $x^m = 1$ and $x^n = 1$; hence $x^{m+n} = x^m \cdot x^n = 1 \cdot 1 = 1$; hence $m + n \in S$. Obviously, $x^0 = 1$; hence

$0 \in S$. Now if $m \in S$, then $x^m = 1$; hence $x^{-m} = x^m \cdot x^{-m} = x^0 = 1$; hence $-m \in S$. This finishes the proof of part (1). Then by part (1) and by the proof of Proposition 2.3, we conclude part (2). □

We note that by choice of m as we have done in the proof of Proposition 2.3, m is the smallest positive integer such that $1, x, x^2, ..., x^{m-1}$ are all distinct. Moreover, any power $x^n \in H$ is equal to one of them. For, by division with remainder, we have

$$n = mq + r,$$

with unique q and r with $0 \leq r < m$; hence

$$x^n = x^{mq} \cdot x^r = (x^m)^q x^r = 1 \cdot x^r.$$

Thus, we have proved that

LEMMA 2.11. *If a cyclic subgroup H of a group G generated by an element x is* not *infinite, then H is a finite group consisting of m element, where m is the smallest positive integer such that $x^m = 1$:*

$$H = \{1, x, x^2, ..., x^{m-1}\},$$

where all the powers are distinct and $x^m = 1$. □

REMARK 2.9. *The obvious difficulty with this lemma is that it is not always possible to verify the hypothesis of this lemma unless the group G itself is a finite set.*

DEFINITION 2.9. *Such a group H is called a* **cyclic group of order** *m.*

The number, i.e., the cardinality, of elements of a group G is called the **order** of G, and is denoted by

$$|G| = \text{ the number of elements of } G.$$

Thus the order of a group is either infinite or finite.

DEFINITION 2.10. *An element x of a group G is said to have* **order** *m (possibly infinite) if the cyclic group it generates has order m.*

This exactly means that m is the smallest positive integer such that $x^m = 1$ or, if the order is infinite, that $x^m \neq 1$ for each $m \neq 0$.

DEFINITION 2.11. *Let U be a subset of a group G. We say that U* ***generates*** *a subgroup H of G if each element of H is expressed as a product of a string of elements of U and of their inverses. If so, we say that the subgroup H is* ***generated*** *by a subset U of G.*

Obviously, the subgroup H of G generated by U is the smallest subgroup of G containing the subset U.

LEMMA 2.12. *A group G of order 2 is cyclic.*

PROOF. G consists of two elements: $G = \{1, x\}$ with $x \neq 1$; hence $x^2 \in G$ must be equal to 1. □

LEMMA 2.13. *A group G of order 3 is cyclic.*

PROOF. **Case 1**: G has one generator $x \neq 1$. In this case, G contains an element x other than 1, and its powers must be all of G but $x^2 \neq x$, so G contains $1, x$, and x^2. Then x^3 can not be equal to x, and x^3 can not be equal to x^3; consequently x^3 must be equal to 1. Thus, G is a cyclic group of order 3 in this case.

Case 2: G has two generators $x \neq 1, y \neq 1$. Then G already contains three elements, and hence xy must be equal to one of these three elements. But, $xy \neq x, xy \neq y$, and hence $xy = 1$. But this means that $y = x^{-1}$ so that y can not be one of the two generators of G. This means that Case 2 does not happen. □

LEMMA 2.14. *A group G of order 4 is either cyclic or the Klein four group* $= \{1, x, y, xy : x^2 = 1, y^2 = 1, xy = yx\}$.

PROOF. **Case 1**: G has a single generator $x \neq 1$. In this case, G is a cyclic of order 4, and $x^4 = 1$.

Case 2: G has two generators $x \neq 1, y \neq 1$. Then, G contains $1, x, y, xy$. Thus, x^2 must be equal to 1, and y^2 must be equal to 1. Then $yx \in G$ must be equal to xy, otherwise, G is not generated by two elements x, y. Thus, G is the Klein 4 group in this case, and it can not be cyclic, although it is an abelian group. □

REMARK 2.10. *What will happen if the order of a group G is ≥ 5?*

If G is a group of order 5, then the case analysis we have naively applied does not work. Fortunately, we have the Lagrange's Theorem (Corollary 2.8), which tells us that G is a cyclic group because its order 5 is a prime number.

If the order of G is 6, then it becomes much more difficult than one might think of. For this, see Proposition 4.6.

EXAMPLE 2.11. *The matrix*

$$\begin{bmatrix} 1 & 1 \\ -1 & 0 \end{bmatrix}$$

is an element of order 6 in $GL_2(\mathbb{R})$, so the cyclic subgroup it generates has order 6. On the other hand, the matrix

$$\begin{bmatrix} 1 & 1 \\ 0 & 1 \end{bmatrix}$$

has infinite order, because we have for each n,

$$\begin{bmatrix} 1 & 1 \\ 0 & 1 \end{bmatrix}^n = \begin{bmatrix} 1 & n \\ 0 & 1 \end{bmatrix}.$$

EXAMPLE 2.12. *In Lemma 2.5, we have seen that the symmetric group S_3 is generated by two permutations x, y.*

EXAMPLE 2.13. *We recall that an $n \times n$ matrix is invertible if and only if its elementary row echelon form is the identity matrix. Thus, this exactly means that GL_n is generated by the elementary matrices.*

EXAMPLE 2.14. *Let V be a subgroup of GL_2 consisting of the four invertible matrices:*

$$\begin{bmatrix} \pm 1 & 0 \\ 0 & \pm 1 \end{bmatrix}.$$

Then one can easily check that

$$x = \begin{bmatrix} 1 & 0 \\ 0 & -1 \end{bmatrix}, \ y = \begin{bmatrix} -1 & 0 \\ 0 & 1 \end{bmatrix}$$

generate V. Thus, any group of order 4 of the form,

$$V = \{1, x, y, xy = yx\},$$

*called the **Klein four group**.*

It is an abelian group which is not cyclic. In fact, it is the smallest group which is not cyclic, because we will later see, as a corollary of Lagrange's Theorem (Corollary 2.8), that any group of prime order is cyclic.

EXAMPLE 2.15. *Let*

$$Q_8 = \{\pm\mathbf{1}, \pm\mathbf{i}, \pm\mathbf{j}, \pm\mathbf{k}\},$$

where

$$\mathbf{1} = \begin{bmatrix} 1 & 0 \\ 0 & 1 \end{bmatrix}, \ \mathbf{i} = \begin{bmatrix} i & 0 \\ 0 & -i \end{bmatrix}, \ \mathbf{j} = \begin{bmatrix} 0 & 1 \\ -1 & 0 \end{bmatrix}, \ \textit{and } \mathbf{k} = \begin{bmatrix} 0 & i \\ i & 0 \end{bmatrix},$$

*where $i = \sqrt{-1}$. Then one can easily check that Q_8 is a subgroup of $GL_2(\mathbb{C})$ of order 8, called the **quaternion group**. Moreover, it is generated by $\mathbf{i}$ and $\mathbf{j}$ and they satisfy the following relations:*

$$\mathbf{i}^4 = \mathbf{1}, \ \mathbf{i}^2 = \mathbf{j}^2, \ \mathbf{ji} = \mathbf{i}^3\mathbf{j} = -\mathbf{k}, \ \mathbf{ij} = \mathbf{k}.$$

Thus, these products should determine the multiplication table of Q_8, and it is not a cyclic group.

REMARK 2.11. *There are in total 5 non-isomorphic groups of order 8, of which three are abelian groups and the rest two are the quarternion group Q_8 and the dihedral group D_4 (Definition 3.6).*

Exercises for Chapter 2.3

EXERCISE 2.15. *List all possible generators of the cyclic group $\mathbb{Z}_6$.*

EXERCISE 2.16. *Show that* $\mathbb{Z}_5$ *is generated by any non-identity element.*

EXERCISE 2.17. *Show that* $\mathbb{Z}_{14}$ *is generated by any one of six generators.*

EXERCISE 2.18. *Make a multiplicative table for the quaternion group.*

EXERCISE 2.19. *Let* G *be a cyclic group of order* n *and let* d *be a divisor of* n*. Prove that* G *has a subgroup of order* d*.*

EXERCISE 2.20. *Figure out all the subgroups of* S_3 *from its multiplication table in (2.2).*

EXERCISE 2.21. *Show that the additive group* $\mathbb{Z}$ *of integers is generated by exactly two elements.*

EXERCISE 2.22. *Prove that a non-empty subset* H *of a group* G *is a subgroup of* G *if and only if* $xy^{-1} \in H$ *whenever* $x, y \in H$*.*

EXERCISE 2.23. *Prove that the* n*-th roots of unity, i.e., the roots of* $x^n = 1$ *is a subgroup of order* n *of the multiplicative group* $\mathbb{C}^{\times}$*.*

EXERCISE 2.24. *Describe all groups which contain no proper nontrivial subgroup.*

EXERCISE 2.25. *Prove that in any group the orders of* ab *and* ba *are the same.*

EXERCISE 2.26. *Prove that the elements of finite order in an abelian group form a subgroup.*

EXERCISE 2.27. *Prove that every subgroup of a cyclic group is cyclic.*

EXERCISE 2.28. *Prove that a group where every non-identity element has order* 2 *is abelian.*

EXERCISE 2.29. *Show by an example the elements of finite order in a non-abelian group do not form a subgroup.*

EXERCISE 2.30. *Prove that a group with no proper subgroup must be cyclic of order* 1 *or prime.*

4. Isomorphisms

Let G and G' be two groups. Although they are different with different operations, if all properties of the group structure of G hold for G' as well, and conversely, they are considered to be the same in our business of classifying mathematical objects, and if this happens we say that they are *isomorphic* to each other.

REMARK 2.12. *This observation is in fact very important in pursuing our better understanding of mathematical objects in general: Suppose we are given two topological spaces X and Y, and we want to know whether they are "topologically" isomorphic, i.e., they share the same topological properties. One way of studying this is to construct a kind of mapping, called a functor F from the category of topological spaces to the category of algebraic spaces, and to use a very famous theorem in Algebraic Topology: If the two topological objects X and Y are topologically isomorphic, then the associated objects $F(X)$ and $F(Y)$ are (algebraically) isomorphic, by trying to find whether there exists a functor F such that $F(X)$ is not (algebraically) isomorphic to $F(Y)$. Once the existence of such a functor is established, by this theorem X can not be topologically isomorphic to Y.*

A typical example of such a kind is: A sphere is not topologically isomorphic to a torus, because their associated algebraic objects (abelian groups) are not (algebraically) isomorphic.

Unlike other centuries, we are currently living in a century where we can perform lots of computational experiments, even using computers, called Algebraic Analysis (i.e., Exact Computation or Symbolic Computation), as long as the original question is mathematically meaningfully converted to a question in Algebra.

For this reason, we will define the concept of being isomorphic on groups. Before that, we have a stimulating example:

EXAMPLE 2.16. *Let G be the set of real upper triangular matrices of the form*

$$\begin{bmatrix} 1 & x \\ 0 & 1 \end{bmatrix}.$$

This is a subgroup of $GL_2(\mathbb{R})$, and the product of two such matrices is

$$\begin{bmatrix} 1 & x \\ 0 & 1 \end{bmatrix}\begin{bmatrix} 1 & y \\ 0 & 1 \end{bmatrix} = \begin{bmatrix} 1 & x+y \\ 0 & 1 \end{bmatrix}.$$

If we examine the upper right entries of the matrices, we see that, whenever such two matrices are multiplied, the two upper right entries x and y are added and all other entries are fixed, and no more than that or no less than that. This means that when we compute out such matrices, we need to keep track of only the upper right entry. This fact is expressed formally by saying that the group G is isomorphic to the additive group of real numbers, and the study on the additive group of real numbers is exactly the same as the study of the subgroup G.

It turns out by a group of French mathematicians, called *Bourbaki*, during the mid 20th century that a right way for this is to relate two groups by a "one-to-one correspondence", first of all, between their elements, and secondly, which is compatible with the laws of composition.

DEFINITION 2.12. *We say that a one-to-one correspondence is* ***compatible with the laws of composition*** *if it satisfies the following property: If $a, b \in G$ correspond to $a', b' \in G'$, respectively, then the product ab in G corresponds to the product $a'b'$ in G'. Any groups G and G' satisfying this property are called* ***isomorphic****.*

When this happens, all properties of the group structure carry over from one group to the other. As examples, we have:

LEMMA 2.15. *The identity elements in isomorphic groups G and G' correspond.*

PROOF. Let the identity element 1 of G corresponds to an element e' in G'. Let a' be an arbitrary element of G' and let a be the corresponding one of G. Since $1a = a$ holds in G, $e'a' = a'$ holds in G', and this shows that e' is in fact the identity element $1'$ of G'. □

Here is one more example:

LEMMA 2.16. *Let G and G' be isomorphic groups. Then the orders of corresponding elements are the same.*

PROOF. Let $a \in G$ correspond to $a' \in G'$. Then since the correspondence is compatible with multiplication, $a^r = 1$ if and only if $a'^r = 1'$. □

Since two isomorphic groups have the same properties, it is often convenient to identify them with each other when speaking informally.

Since a "one-to-one correspondence" is exactly the same as a bijective map, we have a formal definition of being isomorphic:

DEFINITION 2.13. *An* ***isomorphism*** *φ from a group G to another group G' is a bijective map which is compatible with the laws of composition.*

If we write out what this compatibility means using function notation for φ, we get the condition

$$\varphi(ab) = \varphi(a)\varphi(b), \tag{2.3}$$

for each $a, b \in G$.

REMARK 2.13. *Here, we note that the left side of this equality means to multiply a and b in G first and then we apply φ, while on the right side the elements $\varphi(a)$ and $\varphi(b)$, which we denoted by a' and b' before, are multiplied in G'.*

We could also write this condition as

$$(ab)' = a'b'.$$

We also note that the choice of G as domain for this isomorphism is arbitrary: The inverse function $\varphi^{-1} : G' \longrightarrow G$ would serve just as well.

DEFINITION 2.14. *The two groups G and G' are* ***isomorphic*** *if there exists an isomorphism $\varphi : G \longrightarrow G'$.*

NOTATION 2.11. *We will sometimes indicate that two groups are isomorphic by the symbol $\approx$:*

$$G \approx G' \qquad \text{means } G \text{ is isomorphic to } G'.$$

EXAMPLE 2.17. *We can easily verify that the symmetric group S_n of permutations of $\{1, 2, ..., n\}$ is isomorphic to the group of permutations matrices, a subgroup of $GL_n(\mathbb{R})$, and hence we often blur the distinction between these two groups.*

EXAMPLE 2.18. *Let $C = \{..., a^{-2}, a^{-1}, 1, a, a^2, ...\}$ be an infinite cyclic group. Then the map*

$$\varphi : \mathbb{Z}^+ \longrightarrow C$$

defined by sending $n \longmapsto a^n$ is an isomorphism. We note that since the notation is additive in the domain and multiplicative in the range, the condition (2.3) is translated in this case to

$$\varphi(m + n) = \varphi(m)\varphi(n),$$

or

$$a^{m+m} = a^m a^n.$$

This means that studying the additive group $\mathbb{Z}^+$ is sufficiently good enough for studying cyclic groups.

Here is one more simple example:

EXAMPLE 2.19. *Let $G = \{1, x, x^2, ..., x^{n-1}\}$ and $G' = \{1, y, y^2, ..., y^{n-1}\}$ be two cyclic groups, generated by elements x and y of the same order. Then the map sending the generator x to the another generator y is an isomorphism.*

This proves that two cyclic groups of the same order are isomorphic.

DEFINITION 2.15. *The* ***isomorphism class*** *of a group G is the collection of groups which are isomorphic to G.*

Thus, when we speak of **classifying** groups, what is meant is to describe the isomorphism classes. This is generally too difficult to do for all groups, but we will see later that there is only one isomorphism class of groups of order 3 (a corollary of Lagrange's Theorem (Corollary 2.8)), there are two classes of groups of order 4 (Lemma 2.14), and that there are two classes of groups of order 6 (a corollary of the First Sylow Theorem, Proposition 4.6) and five classes of groups of order 12 (a corollary of the Three Sylow Theorems, Proposition 4.9).

DEFINITION 2.16. *Let G be a group. An* ***automorphism*** *of G is an isomorphism from G to itself.*

The identity map is an automorphism, but there are nearly always other automorphisms as well.

EXAMPLE 2.20. *Let $G = \{1, x, x^2\}$ be a cyclic group of order 3. That is, x is a generator of G of order 3, so that $x^3 = 1$. Then the transposition with interchanges x and x^2 is an automorphism of G:*

$$\begin{aligned} 1 &\mapsto 1 \\ x &\mapsto x^2 \\ x^2 &\mapsto x. \end{aligned}$$

This is because x^2 is another generator for G of order 3. That is, if we call this element y, then $\{1, y, y^2\}$ generated by y is the group G, and this automorphism compares the two realizations of G as a cyclic group.

The most important example of automorphism is conjugation:

DEFINITION 2.17. *Let b be an element of a group G.* ***Conjugation by*** *b is the map φ from G to itself defined by*

$$\varphi(x) = bxb^{-1}.$$

LEMMA 2.17. *Conjugation is an automorphism.*

PROOF. Conjugation is compatible with multiplication in the group, because

$$\varphi(xy) = bxyb^{-1} = bxb^{-1}byb^{-1} = \varphi(x)\varphi(y).$$

Conjugation is bijective, because it has an inverse function, that is, conjugation by b^{-1}. □

REMARK 2.14. *We note that conjugation is the identity map if the group G is abelian. But, any non-abelian group always has some non-trivial conjugation, and hence it has non-trivial automorphism.*

DEFINITION 2.18. *The element bab^{-1} of a group G is called the* ***conjugate of*** a ***by*** b. *Two elements a, a' of a group G are called* ***conjugate*** *if $a' = bab^{-1}$ for some $b \in G$.*

The conjugate of a behaves in much the same way as the element a itself. For example, the conjugate has the same order in the group. This is because it is the image of a by an automorphism.

Moreover, if we write bab^{-1} by a', then

$$ba = a'b,$$

so that we can think of conjugation by b as the change in a which results when one moves b from one side in ba to the other. Thus, the conjugate has a useful interpretation, though trivial.

Exercises for Chapter 2.4

EXERCISE 2.31. *Prove that the inverse function of an isomorphism of groups is an isomorphism.*

EXERCISE 2.32. *Show that the additive group $\mathbb{Z}_6$ is isomorphic to $\mathbb{Z}_2 \times \mathbb{Z}_3$.*

EXERCISE 2.33. *Show that the Klein 4 group V is not isomorphic to $\mathbb{Z}_4$.*

EXERCISE 2.34. *Determine all groups of order 4 up to isomorphism.*

EXERCISE 2.35. *Show that* $\mathbb{Z}_3$ *has two automorphisms and prove that* $Aut(\mathbb{Z}_3) \simeq \mathbb{Z}_2$.

EXERCISE 2.36. *Show that the Klein* 4 *group* $\mathbb{Z}_2 \times \mathbb{Z}_2$ *has six automorphisms and prove that* $Aut(\mathbb{Z}_2 \times \mathbb{Z}_2) \simeq S_3$.

EXERCISE 2.37. *Prove that* $Aut(D_3) \simeq D_3$.

EXERCISE 2.38. *Show that the quaternion group in Example 2.15 is not isomorphic to* D_4.

EXERCISE 2.39. *Prove that the products* ab *and* ba *are conjugate elements in a group.*

EXERCISE 2.40. *Show that the matrices*

$$\begin{bmatrix} 1 & 1 \\ 0 & 1 \end{bmatrix}, \begin{bmatrix} 1 & 0 \\ 1 & 1 \end{bmatrix}$$

are conjugate elements in $GL_2(\mathbb{R})$, *but that they are not conjugate in* $SL_2(\mathbb{R})$.

EXERCISE 2.41. *Determine the group of automorphisms of a group* G, *when* G *is* $\mathbb{Z}^+$, *a cyclic group of order* 10, *or* S_3.

5. Homomorphisms

As is the case of sets, we compare two groups in terms of maps which are compatible with the laws of composition, called homomorphisms of groups:

DEFINITION 2.19. *Let* G, G' *be groups. A* ***homomorphism*** $\varphi : G \longrightarrow G'$ ***of groups***, *or a* ***group homomorphism*** $\varphi : G \longrightarrow G'$, *is a map satisfying the rule*

$$\varphi(ab) = \varphi(a)\varphi(b)$$

for each $a, b \in G$. *In particular, if* $G = G'$, *then the homomorphism* φ *is called an* ***endomorphism*** *of* G. *If, in particular, an endomorphism is an isomorphism, then it is called an* ***automorphism*** *of* G.

We have learned that an isomorphism also satisfies this condition.

EXAMPLE 2.21. *Here are some examples of homomorphisms:*

(1) *the determinant function,* $\det : GL_n(\mathbb{R}) \longrightarrow \mathbb{R}^\times$*;*
(2) *the sign of a permutation, sign :* $S_n \longrightarrow \{\pm 1\}$*, where the sign of a permutation* σ *is, by definition, the determinant of the corresponding permutation matrix of* σ*.*
(3) *the map* $\varphi : \mathbb{Z}^+ \longrightarrow G$ *defined by* $\varphi(n) = a^n$*, where* a *is a fixed element of* G*;*
(4) *the* ***inclusion map*** $i : H \longrightarrow G$ *of a subgroup* H *into a group* G*, defined by* $i(x) = x$*.*

A homomorphism of groups is very special in the sense that it carries the identity element to the identity element:

PROPOSITION 2.6. *A group homomorphism* $\varphi : G \longrightarrow G'$ *carries the identity element to the identity element, and inverses to inverses. Namely,*

$$\varphi(1_G) = 1_{G'}, \qquad \textit{and} \qquad \varphi(a^{-1}) = \varphi(a)^{-1}.$$

PROOF. Since φ is a homomorphism, we have $\varphi(1) = \varphi(1 \cdot 1) = \varphi(1)\varphi(1)$; hence cancelling out $\varphi(1)$ from both sides of this equality, we get $1 = \varphi(1)$. Now, $1 = \varphi(1) = \varphi(aa^{-1}) = \varphi(a)\varphi(a^{-1})$, and similarly $1 = \varphi(a^{-1})\varphi(a)$. Hence, by Lemma 2.1, we conclude that $\varphi(a^{-1}) = \varphi(a)^{-1}$, the unique inverse of $\varphi(a)$. □

NOTATION 2.12. *When we write the identity elements* $1_G, 1_{G'}$ *of groups* G*,* G'*, respectively, we usually drop the subscripts* G, G'*, when there is no danger of confusion, which is mostly the case.*

We should note that whenever a homomorphism of groups

$$\varphi : G \longrightarrow G',$$

there are two naturally associated subgroups, the kernel of φ and the image of φ. Such associations should be automatic to everyone.

DEFINITION 2.20. *Let $\varphi : G \longrightarrow G'$ be a homomorphism of groups. Define two subsets, the **kernel** of φ and the **image** of φ, of the domain G and the range G', respectively, as follows:*

$$\begin{aligned} im(\varphi) &:= \{y \in G' : y = \varphi(x) \text{ for some } x \in G\}, \\ \ker(\varphi) &:= \varphi^{-1}(1), \end{aligned}$$

where $\varphi^{-1}(1)$ is the inverse image of the identity element of G':

$$\varphi^{-1}(1) := \{x \in G : \varphi(x) = 1\}.$$

REMARK 2.15. *In categorical language, the "dual" of the sub-object, called the kernel, is not the image, but the quotient object, the "cokernel". But, the definition of the cokernel of a homomorphism of groups requires first what a quotient group means, and this new concept will be presented in a later section, called Quotient Groups, of this Chapter.*

LEMMA 2.18. *Let $\varphi : G \longrightarrow G'$ be a homomorphism of groups. Then $im(\varphi)$ is a subgroup of G', called the image of φ.*

PROOF. We have shown in the above proposition that $\varphi(1) = 1$. Thus, the identity element 1 of G' is an element of $im(\varphi)$. Now, if $y, z \in im(\varphi)$, then $y = \varphi(a)$ and $z = \varphi(b)$ for some $a, b \in G$; hence $yz = \varphi(a)\varphi(b) = \varphi(ab)$, that is, $yz \in im(\varphi)$. If $y = \varphi(a) \in im(\varphi)$, then $y^{-1} = \varphi(a)^{-1} = \varphi(a^{-1})$, that is, $y^{-1} \in im(\varphi)$. □

EXAMPLE 2.22. *This is a continuation of Example 2.21:*

(1) *The image of the determinant function* $\det : GL_n(\mathbb{R}) \longrightarrow \mathbb{R}^\times$ *is equal to the range of the map, i.e., the determinant function is surjective.*
(2) *The image of the sign of a permutation, sign : $S_n \longrightarrow \{\pm 1\}$ is equal to the range of the map, i.e., the sign map is surjective.*
(3) *The image of the map $\varphi : \mathbb{Z}^+ \longrightarrow G$ defined by $\varphi(n) = a^n$, where a is a fixed element of G, is a cyclic subgroup of G generated by a.*

(4) *The image of the inclusion map* $i : H \longrightarrow G$ *of a subgroup* H *into a group* G*, defined by* $i(x) = x$*, is the subgroup* H*.*

LEMMA 2.19. *Let* $\varphi : G \longrightarrow G'$ *be a homomorphism of groups. Then* $\ker(\varphi)$ *is a subgroup of* G*, called the* ***kernel*** *of* φ *.*

PROOF. We have shown in the above proposition that $\varphi(1) = 1$. Thus, the identity element 1 of G is an element of $\ker(\varphi)$. Now, if $a, b \in \ker(\varphi)$, then $\varphi(a) = 1 = \varphi(b)$; hence $\varphi(ab) = \varphi(a)\varphi(b) = 1 \cdot 1 = 1$, that is, $ab \in \ker(\varphi)$. If $a \in \ker(\varphi)$, then $\varphi(a) = 1$; hence $1 = \varphi(a)^{-1} = \varphi(a^{-1})$, that is, $a^{-1} \in G$. □

EXAMPLE 2.23. *This is a continuation of Example 2.21.*

(1) *The kernel of the determinant homomorphism is the subgroup of matrices whose determinant is equal to 1. This subgroup is denoted by* $SL_n(\mathbb{R})$*:*

$$SL_n(\mathbb{R}) = \{A \in GL_n(\mathbb{R}) : \det(A) = 1\},$$

and is called the ***special linear group****.*

(2) *The kernel of the sign function, sign* : $S_n \longrightarrow \{\pm 1\}$*, is called the* ***alternating group****. This subgroup of* S_n *is denoted by* A_n*:*

$$A_n = \{\textit{even permutations of } \{1, 2, ..., n\}\}.$$

(3) *The kernel of the map* $\varphi : \mathbb{Z}^+ \longrightarrow G$ *defined by* $\varphi(n) = a^n$*, where* a *is a fixed element of* G*, is the set of integers* n *such that* $a^n = 1$*. This is a subgroup of* $\mathbb{Z}^+$ *by Lemma 2.10.*

In addition to be a subgroup, the kernel of a homomorphism has an extra property which is subtle but very important. Here is what it is:

LEMMA 2.20. *Let* $\varphi : G \longrightarrow G'$ *be a homomorphism of groups. If* $a \in \ker(\varphi)$*, then for any element* b *of* G*,*

$$bab^{-1} \in \ker(\varphi).$$

PROOF. This is because $\varphi(bab^{-1}) = \varphi(b)\varphi(a)\varphi(b^{-1}) = \varphi(b) \cdot 1 \cdot \varphi(b^{-1}) = \varphi(b)\varphi(b)^{-1} = 1$. □

DEFINITION 2.21. *A subgroup N of a group G is called a* ***normal subgroup*** *if it satisfies the following property: For each $a \in N$ and for each $b \in G$, the conjugate bab^{-1} is an element of N.*

In the above, we have already proven that

COROLLARY 2.5. *The kernel of a homomorphism of groups $G \longrightarrow G'$ is a normal subgroup of G.*

EXAMPLE 2.24. *Thus,*

(1) *$SL_n(\mathbb{R})$ is a normal subgroup of $GL_n(\mathbb{R})$;*
(2) *A_n is a normal subgroup of S_n.*

We note that *any subgroup of an abelian group is normal*, because in this case $bab^{-1} = abb^{-1} = a$.
But subgroups of non-abelian groups need not be normal in general.

EXAMPLE 2.25. *Let T be the set of all invertible upper triangular real matrices of the same size n. Then T is a subgroup of $GL_n(\mathbb{R})$, but it is not a normal subgroup of $GL_n(\mathbb{R})$ in general. For this, we take $n = 2$ and consider two matrices*

$$A = \begin{bmatrix} 1 & 1 \\ 0 & 1 \end{bmatrix}, \quad B = \begin{bmatrix} 0 & 1 \\ 1 & 0 \end{bmatrix}.$$

Then $A \in T$ and $B \in GL_2(\mathbb{R})$, but

$$BAB^{-1} = \begin{bmatrix} 1 & 0 \\ 1 & 1 \end{bmatrix} \notin T.$$

REMARK 2.16. *At the end of a later Section of this Chapter, called Cosets and Normal Subgroups, we will have a proposition (Proposition 2.10) which tells us more about the normality of a subgroup H of a group G, and this proposition should provide us with a lot better understanding on the normality of a subgroup.*

There are normal subgroups which are moreover abelian:

DEFINITION 2.22. *The* **center** *of a group G, denoted by Z or $Z(G)$, is the set of elements of G which commutes with each element of G:*

$$Z = \{z \in G : zx = xz \text{ for each } x \in G\}.$$

LEMMA 2.21. *The center Z of a group G is an abelian subgroup of G.*

PROOF. Obviously, the identity element 1 of G belongs to Z. Now, if $a, b \in Z$, then for any $x \in G$,

$$abx = a(bx) = a(xb) = (ax)b = (xa)b = xab,$$

that is, $ab \in Z$. If $a \in Z$, then for any $x \in G$,

$$a^{-1}x = a^{-1}(x^{-1})^{-1} = (x^{-1}a)^{-1} = (ax^{-1})^{-1} = xa^{-1},$$

that is, $a^{-1} \in Z$. Thus Z is a subgroup of G. Then by definition, the center is obviously abelian. □

LEMMA 2.22. *A group G is abelian if and only if its center is equal to G.*

PROOF. If G is abelian, then its center is obviously G itself. If G is not abelian, then there exist elements $y, x \in G$ such that $yx \neq xy$. Thus, this $y \notin Z$; hence the center Z is a proper subgroup of G. □

EXAMPLE 2.26. *By Corollary 2.2, we know that S_3 is the non-abelian group of least order. Then the relations between two generators x, y of S_3,*

$$x^3 = 1, y^2 = 1, x^2y = yx,$$

clearly tells us that the center of S_3 is the identity subgroup $\{1\}$.

LEMMA 2.23. *The center Z of a group G is a normal subgroup of G.*

PROOF. Let $z \in Z$ and x be any element of G. Then

$$xzx^{-1} = zxx^{-1} = z \in Z.$$

□

REMARK 2.17. *We have seen that the center of a group G is the largest abelian subgroup of G, you can think of, that is also normal, and it sometimes plays an important role when we study the structure of non-abelian groups.*

Exercises for Chapter 2.5

EXERCISE 2.42. *Show that $\mathbb{Z}_n$ has exactly n endomorphisms.*

EXERCISE 2.43. *Fine all subgroups of S_3, and determine which are normal.*

EXERCISE 2.44. *Fine all subgroups of the quaternion group, and determine which are normal.*

EXERCISE 2.45. *Let $\varphi : G \longrightarrow G'$ be a group homomorphism. Prove that $\varphi(x) = \varphi(y)$ if and only if $xy^{-1} \in \ker(\varphi)$.*

EXERCISE 2.46. *Prove that the center of $GL_n(\mathbb{R})$ is the group of **scalar matrices**, that is, those of the form cI, $c \neq 0 \in \mathbb{R}$.*

EXERCISE 2.47. *Which symmetric groups and which alternating groups are abelian?*

EXERCISE 2.48. *Show that two conjugate permutations σ and $\tau\sigma\tau^{-1}$ have the same sign but not necessarily the same number of transpositions.*

EXERCISE 2.49. *Show by providing an example that $GL_n(\mathbb{R})$ is not a normal subgroup of $GL_n(\mathbb{C})$.*

EXERCISE 2.50. *Let $\varphi : G \longrightarrow G'$ be a surjective homomorphism of groups.*

(1) *Prove that if G is cyclic then so is G'.*
(2) *Prove that if G is abelian then so is G'.*
(3) *Prove that H is a normal subgroup of G then so is $\varphi(H)$ in G'.*

6. Equivalence Relations and Partitions

In this section, we are going to review a fundamental mathematical construction in terms of a property P. Given a set S and a property P, we form a new set, a subset of S, by gathering those elements of S satisfying the property P. It is usually the case that the property P is given as a sort of equating certain elements of S, according to a given rule.

EXAMPLE 2.27. *Let S be the set $\mathbb{Z}$ of all integers and $P(x)$ denote a statement with unknown x:*

$$P(x) = \text{ “}x \text{ is an integer multiple of } 2\text{”}.$$

Then,

$$H = \{x \in \mathbb{Z} : P(x)\}$$

consists of all even integers. It is a non-empty subset of $\mathbb{Z}$. Obviously then, the set of all integers is equal to a disjoint union of H and its complement, H^c, the set of all odd integers:

$$\mathbb{Z} = H \cup H^c, \qquad H \cap H^c = \varnothing.$$

Thus, any integer belongs to either one of these subsets, but not both.

EXAMPLE 2.28. *Let S be the set of triangles in the plane. When we view “congruent” triangles in the plane as equivalent geometric objects, S is a disjoint union of the subsets consisting of “congruent” triangles.*

In what follows, we are going to abstract these procedures formally to an important notion:

DEFINITION 2.23. *Let S be a set. A* **partition** *P of S is a subdivision of S into non-overlapping subsets:*

$$S = \text{ the union of disjoint and non-empty subsets.}$$

EXAMPLE 2.29. *The sets*

$$\{1, 3, 5\}, \{2, 4\},$$

form a partition of the set $\{1, 2, 3, 4, 5\}$.

EXAMPLE 2.30. *The two sets, of even natural numbers and of odd natural numbers, form a partition of the set* $\mathbb{N}$ *of all natural numbers.*

Another important notion, which is in fact logically equivalent to the notion of a partition of a set S, is the following one:

DEFINITION 2.24. *An* ***equivalence relation****, denoted by*

$$\sim$$

on a set S *is a (binary) relation satisfying the following conditions: for each* $a, b, c \in S$*,*

(1) *(transitive) If* $a \sim b$ *and* $b \sim c$*, then* $a \sim c$*;*
(2) *(symmetric) If* $a \sim b$*, then* $b \sim a$*;*
(3) *(reflexive)* $a \sim a$*.*

EXAMPLE 2.31. *Congruence of triangles in the plane is an example of an equivalence relation on the set* S *of all triangles in the plane.*

EXAMPLE 2.32. *For any integers* a, b*, define* $a \sim b$ *if* $a - b$ *is even. Then this relation is definitely an equivalence relation on the set* $\mathbb{Z}$ *of all integers.*

For, if $a \sim b$ *and* $b \sim c$*, then* $a - b$ *and* $b - c$ *are both even, and hence their sum* $a - b$ *is also even; if* $a - b$ *is even then obviously,* $b - a$ *is even; finally,* $a - a = 0 \cdot 2$ *is also even.*

LEMMA 2.24. *The two notions of a partition of a set* S *and an equivalence relation on* S *are logically equivalent.*

PROOF. Suppose that a partition P on a set S is given. Then we define an equivalence relation E by the rule $a \sim b$ if both a and b lie in the same subset of the partition. Then the axioms (i.e., conditions) for a relation to be an equivalence relation defined in Definition 2.24 are obviously satisfied.

Conversely, given an equivalence relation E on a set S, we define a partition P of S as follows: The subset containing a is the set of all elements b such that $a \sim b$. This subset is called the **equivalence class** of a, and then S is partitioned into equivalence classes, as we shall see it now:

Let C_a denote the equivalence class of a:

$$C_a = \{b \in S : a \sim b\}.$$

Then as a runs through the set S, these subsets C_a surely cover the set S, that is, their union is equal to S. Thus, the collection of C_a forms a partition of S, provided that C_a and C_b are the same if $C_a \cap C_b \neq \varnothing$, i.e., if they have an element in common. Thus, we will finish our proof if we can prove the following statement:

(2.4) If $C_a \cap C_b \neq \varnothing$, then $C_a = C_b$.

For this, suppose that $d \in C_a \cap C_b$. Then $a \sim d$ and $b \sim d$.

Thus, if we can show that $a \sim d$ implies $C_a = C_d$, then we are done.

For this, suppose that $a \sim d$. For any $x \in C_d$, $d \sim x$ holds; hence by transitivity of $\sim$, $a \sim x$, that is, $x \in C_a$, and this shows that $C_d \subseteq C_a$. By interchanging the roles of a and d, the same argument also proves that $C_a \subseteq C_d$. □

REMARK 2.18. *The usefulness of the notion of a partition of a set S or equivalently, the notion of an equivalence relation on S, is due to the fact that it allows us to construct another set $\bar{S}$, and it goes as follows: Suppose we are given an equivalence relation of a partition on a set S. Then the set of the equivalence classes is another set, denoted by*

$$\bar{S}.$$

Thus, the elements of $\bar{S}$ make up the partition of S.

NOTATION 2.13. *The equivalence class of a, or the subset of the partition containing a, is denoted by*

$$\bar{a}.$$

Thus, $\bar{a}$ is an element of $\bar{S}$.

In particular, we note that there exists a natural surjective map: $S \longrightarrow \bar{S}$, which sends

$$a \longmapsto \bar{a}.$$

EXAMPLE 2.33. *Let S be the set $\mathbb{Z}$ of all integers, partitioned into the subset $\bar{0}$ of even integers and the subset $\bar{1}$ of odd integers. Thus,*

$$\mathbb{Z} = \bar{0} \cup \bar{1},$$

and hence $\bar{0} = \bar{2} = \bar{4}$ and so on, and thus, we have:

$$\bar{S} = \{\bar{0}, \bar{1}\}.$$

Then, the natural map: $S \longrightarrow \bar{S}$ defined by sending any even integer to $\bar{0}$ and any odd integer to $\bar{1}$ is obviously surjective.

REMARK 2.19. *We note in the above example that $\bar{0}$ is the set of all even integers and $\bar{1}$ is the set of all odd integers, and $\bar{2}$ is also the set of all even integers. Thus, we need to choose a* representative, *i.e., a particular element, for the set of all even integers, and in the above we choose 0 to be such an element of S. Similarly, we chose 1 to be a representative of the set of all odd integers. That is, the notation $\bar{0}$ itself just means that 0 is a representative for the set $\bar{0}$ of all even integers, and we can choose 2 as another representative for the set of all even integers.*

LEMMA 2.25. *Let $\varphi : S \longrightarrow T$ be a map, and define $a \sim b$ if $\varphi(a) = \varphi(b)$. This defines an equivalence relation on the domain S.*

PROOF. This is because the domain S is a disjoint union of the inverse images,

$$\varphi^{-1}(x) = \{a \in G : \varphi(a) = x\},$$

where x runs through the image $im(\varphi)$, i.e., we have a partition of S induced by the map φ. And then, as we have already seen above, any two elements a, b of S are equivalent if they belong to the same subset $\varphi^{-1}(x)$, i.e., their images $\varphi(a)$ and $\varphi(b)$ are the same, and this is what we have now. $\square$

REMARK 2.20. *The equivalence relation on the domain of a map is called the* **equivalence relation determined by the map**, *and the inverse images* $\varphi^{-1}(x)$ *are also called the* **fibres** *of the map* φ.

Thus, the non-empty fibres form a partition on the domain S *of a map. Moreover, we note that the set* $\bar{S}$ *of equivalence classes, which is the set of non-empty fibres, is bijectively mapped to the image,* $im(\varphi)$, *by sending an element* $\bar{s}$ *of* $\bar{S}$ *to* $\varphi(s)$,

$$\bar{s} \longmapsto \varphi(s). \tag{2.5}$$

EXAMPLE 2.34. *Let* $\varphi : \mathbb{C}^\times \longrightarrow \mathbb{R}^\times$ *be the absolute value homomorphism defined by*

$$\varphi(a) = |a|\,.$$

Thus induced equivalence relation on the domain is $a \sim b$ *if* $|a| = |b|$, *and the non-empty fibres of this map are the concentric circles about* 0. *Then, the non-empty fibres are obviously in one-to-one correspondence with element of* $im(\varphi)$, *the set of all positive real numbers.*

REMARK 2.21. *So far, we have not considered what will happen or what we can do more, if both the domain and the range of a map* φ *have the laws of composition, respectively. Our final answer to this is given in a later section, called Quotient Groups.*

For this, we go back to group homomorphisms. Let $\varphi : G \longrightarrow G'$ be a homomorphism of groups, and consider the associated equivalence relation $\sim$ on the domain G.

NOTATION 2.14. *For a historical reason, this equivalence relation* $\sim$ *is usually denoted by* $\equiv$ *rather than by* $\sim$, *and is referred to as* **congruence:**

$$a \equiv b \qquad \text{if} \qquad \varphi(a) = \varphi(b), \tag{2.6}$$

because of the so-called modular arithmetic, which we will later study what it is in a Section, called Modular Arithmetic of this Chapter. We note that it was Gauss (1777 - 1855) who first introduced such a notation $\equiv$ *in his very famous book,* Disquisitiones Arithmeticae, *published in 1801.*

The relation (2.6) can be rewritten in a number of different ways, but of which the following is going to be the most important for us:

PROPOSITION 2.7. *Let $\varphi : G \longrightarrow G'$ be a group homomorphism with kernel N. Let $a, b \in G$. Then $\varphi(a) = \varphi(b)$ if and only if $b = an$ for some element $n \in N$, or equivalently, if $a^{-1}b \in N$, i.e., $b \in aN$.*

PROOF. Suppose that $\varphi(a) = \varphi(b)$. Then

$$1 = \varphi(a)^{-1}\varphi(b) = \varphi(a^{-1})\varphi(b) = \varphi(a^{-1}b).$$

Hence $a^{-1}b \in \ker(\varphi) = N$, that is, $a^{-1}b = n$ for some $n \in N$. Conversely, if $b = an$ for some $n \in N$, then

$$\varphi(b) = \varphi(an) = \varphi(a)\varphi(1) = \varphi(a) \cdot 1 = \varphi(a).$$

□

NOTATION 2.15. *The set of elements of the form, an, is denoted by aN, and is called a (left)* **coset** *of N in G:*

$$aN = \{g \in G : g = an \qquad \textit{for some } n \in N\}.$$

Thus, if $\varphi : G \longrightarrow G'$ is a homomorphism of groups with kernel N, then the coset aN is the set of all group elements b which are congruent to a, and the congruence relation $a \equiv b$ partitions the group G into **congruence classes**, the cosets aN, the non-empty fibres of the map φ.

An important case to look at is when the kernel is the trivial subgroup $\{1\}$:

COROLLARY 2.6. *A group homomorphism $\varphi : G \longrightarrow G'$ is injective if and only if its kernel is the trivial subgroup $\{1\}$.*

PROOF. We recall that φ is injective if and only if $\varphi(a) = \varphi(b)$ implies $a = b$. Thus, if φ is injective, then since any $x \in N$ is mapped to 1, we have $\varphi(x) = 1 = \varphi(1)$; hence $x = 1$, that is, N is the trivial group $\{1\}$. Conversely, if $N = \{1\}$, then by Proposition 2.7 we see that $\varphi(a) = \varphi(b)$ implies $a = b \cdot 1$, and hence φ is injective. □

REMARK 2.22. *Thus, to verify that a homomorphism* $\varphi : G \longrightarrow G'$ *of groups is an isomorphism, we just check whether the following two conditions hold:*

(1) *its kernel is the trivial group, i.e.,* $\ker(\varphi) = \{1\}$*;*
(2) *its image is* G'*, i.e.,* $im(\varphi) = G'$*,*

Thus, if one of these two does not hold, then φ *is not an isomorphism.*

Exercises for Chapter 2.6

EXERCISE 2.51. *Prove that the non-empty fibres of a map form a partition of the domain.*

EXERCISE 2.52. *Prove that* $aN = \varphi^{-1}(\varphi(a)) = Na$*, where* N *is the kernel of a group homomorphism* $\varphi : G \longrightarrow G'$*.*

EXERCISE 2.53. *Let* H *be a subgroup of a group* G*. Prove that the relation defined by* $a \sim b$ *if* $b^{-1}a \in H$ *is an equivalence relation on* G*.*

EXERCISE 2.54. *Prove that the relation* x *conjugate to* y *in a group* G *is an equivalence relation on* G*.*

7. Cosets and Normal Subgroups

In the previous section, we have introduced the notion of cosets of the kernel as equivalence classes when we are given a group homomorphism. In fact, it turns out that we can define a coset of a subgroup H, whenever a subgroup H of a group G is given.

DEFINITION 2.25. *Let* H *be a subgroup of a group* G*. A* **(left) coset** *of* H *is a subset of* G *of the form*

$$aH = \{ah : h \in H\},$$

where a *is an element of* G*.*

Since $H = 1H$, the subgroup H itself is a left coset of H.

LEMMA 2.26. *Let H be a subgroup of a group G. The cosets of H are equivalence classes for the* **congruence** *relation,*

$$a \equiv b \qquad \text{if } b = ah, \text{ for some } h \in H, \text{i.e.}, a^{-1}b \in H.$$

PROOF. We need to verify the three axioms or conditions for this congruence to be an equivalence relation:

(1) (transitivity) Suppose that $a \equiv b$ and $b \equiv c$. Thus, $b = ah$ and $c = bh'$ for some $h, h' \in H$; hence $c = bh' = ahh'$; hence $a \equiv c$.
(2) (symmetry) Suppose that $a \equiv b$. Thus, $b = ah$ for some $h \in H$; hence $a = bh^{-1}$, where $h^{-1} \in H$ because H is a subgroup of G; hence $b \equiv a$.
(3) (reflexivity) $a = a \cdot 1$ and $1 \in H$; hence $a \equiv a$.

□

COROLLARY 2.7. *The left cosets of a subgroup of a group G partition the group G.*

PROOF. This is because equivalence classes form a partition, as we have discussed in the previous section. □

REMARK 2.23. *The notation aH defines a certain subset of the group G. We have already seen that aH is the unique coset containing a, and*

$$aH = bH \text{ if and only if } a \equiv b$$

holds. Thus, a different notation, bH, represents the same subset if and only if $a \equiv b$, and this corollary just restates (2.4):

(2.7) *If aH and bH have a common element, then they are the same.*

EXAMPLE 2.35. *Let G be the symmetric group S_3, whose representation is given in Example (2.1):*

$$G = \{1, x, x^2, y, xy, x^2y\},$$

whose defining relations are:

$$x^3 = 1, y^2 = 1, x^2y = yx.$$

The element xy *has order* 2, *and hence it generates a cyclic subgroup* $H = \{1, xy\}$ *of order* 2. *Then the left cosets of* H *in* G *are the three sets*

$$\begin{aligned} \{1, xy\} &= H = xyH, \\ \{x, x^2y\} &= xH = x^2yH, \\ \{x^2, y\} &= x^2H = yH. \end{aligned} \tag{2.8}$$

We note that they partition S_3.

In particular, we also note that the number of cosets times the order of H *is* 6, *the number of order of* S_3.

By the way, do you think this is a mere coincidence? The answer is no, as the following explains:

DEFINITION 2.26. *Let* H *be a subgroup of a group* G. *The number of left cosets of* H *is called the* ***index*** *of* H *in* G, *and is denoted by*

$$[G : H].$$

Thus, the index of H in S_3 in the above example is 3. Obviously, if G is an infinite group, then the index may be infinite too.

LEMMA 2.27. *Let* H *be a subgroup of a group* G.

(1) *Then each left coset* aH *has the same number of elements as* H *does.*
(2) *(Counting Formula) We have:*

$$|G| = |H| \cdot [G : H], \tag{2.9}$$

where $|G|$ *denotes the order of* G, *and the equality still holds if some terms are infinite.*

PROOF. Since the map $H \longrightarrow aH$ defined by sending

$$h \longmapsto ah$$

is obviously bijective, i.e., one-to-one and onto. This proves Part (1); hence also establishes the equality in Part (2). $\square$

The fact that one of the two terms on the right side of equation (2.9) must divide the left side is very important, and here is one of these conclusions:

COROLLARY 2.8. *(Lagrange's Theorem) Let G be a finite group, and H a subgroup of G. Then the order of H divides the order of G.* □

REMARK 2.24. *The converse of Lagrange's Theorem (Corollary 2.8) does not hold in general. (See Corollary 4.12.)*

In Section 2 we have defined the order of an element $a \in G$ to be the order of the cyclic subgroup generated by a. Thus, Lagrange's Theorem implies:

COROLLARY 2.9. *The order of an element of a group G divides the order of G.* □

This provides us with a remarkable consequence, which allows us to classify all the groups of prime order p:

COROLLARY 2.10. *Suppose that G is a group of order p and moreover p is a prime number. Then any non-identity element a of G generates G as a cyclic group:*

$$G = \{1, a, a^2, ..., a^{p-1}\}.$$

PROOF. Since $a \neq 1$, the order of a is strictly greater than 1, and it should divide $p = |G|$; hence the order of a must be p. Since G has order p, the cyclic subgroup $\{1, a, ..., a^{p-1}\}$ must be the whole group G. □

REMARK 2.25. *Thus, we have classified all the groups of prime order p. They form one isomorphism class, the class of a cyclic group of order p.*

REMARK 2.26. *The next question is what will happen, or what can we say more, when the order of a group is a power of a prime number p.*

It turns out that this question is more complicated than one expects, and the number of possibilities for groups of order p^n increases very rapidly with n.

The first easy case is:

PROPOSITION 2.8. *Every group of order* 4 *is abelian.*

PROOF. Let G be a group of order 4. If there exists an element x of order 4, then G is a cyclic group, generated by this element x; hence it is abelian. Otherwise, by Lagrange's Theorem, G consists of the three distinct non-identity elements, x, y, xy of order 2 and the identity element 1. Then one can easily conclude that $yx = xy$ because, for instance, if $yx = x$ we then have $y = 1$, but this contradicts our choice of y. □

EXAMPLE 2.36. *Here are two more examples of abelian groups whose verifications are omitted:*

(1) There are five isomorphism classes of groups of order 2^3, among them the three are abelian.
(2) There are 14 isomorphism classes of groups of order 2^4, among them the five are abelian.

We also note that the Counting Formula can also be applied when we are given a group homomorphism $\varphi : G \longrightarrow G'$.

COROLLARY 2.11. *Let* $\varphi : G \longrightarrow G'$ *be a homomorphism of groups. Assume that* G *is a* finite *group. Then*

$$|G| = |\ker(\varphi)| \cdot |im(\varphi)| .$$

Thus, $|\ker(\varphi)|$ *divides* $|G|$, *and* $|im(\varphi)|$ *divides both* $|G|$ *and* $|G'|$.

PROOF. Since the number of the left cosets of $\ker(\varphi)$ is equal to the number of the non-empty fibres of the map φ by Proposition 2.7, and since we have already seen in Remark 2.20 that there exists a bijection between the set of the non-empty fibres and $im(\varphi)$, we have an equality:

$$[G : \ker(\varphi)] = |im(\varphi)| .$$

Thus, by (2.9) we obtain,

$$|G| = |\ker(\varphi)| \cdot |im(\varphi)| ,$$

as claimed. This equality implies that $|\ker(\varphi)|$ and $|im(\varphi)|$ divide $|G|$. On the other hand, since $im(\varphi)$ is a subgroup of G', $|im(\varphi)|$ divides $|G'|$ as well, by the Counting Formula again. □

So far, we have worked with left cosets aH. We can also define right cosets of a subgroup H of a group G and repeat the above discussion for them.

DEFINITION 2.27. *A **right coset** of a subgroup H of a group G is a subset of G of the form:*

$$Ha = \{ha : h \in H\},$$

where a is an element of G.

Since $H = H1$, the subgroup H itself is a right coset of H. The right cosets Ha are equivalence classes for the relation (called **right congruence**)

$$a \equiv b \text{ if } b = ha, \text{ for some } h \in H.$$

REMARK 2.27. *We note that right cosets need not be the same as left cosets.*

EXAMPLE 2.37. *Let G be the symmetric group S_3 and $H = \{1, xy\}$, as before. Then the right cosets of H in S_3 are:*

$$\begin{aligned} \{1, xy\} &= H = Hxy, \\ \{x, y\} &= Hx = Hy, \\ \{x^2, x^2y\} &= Hx^2 = Hx^2y. \end{aligned} \tag{2.10}$$

This partition of S_3 is not the same as the partition of S_3 into the left cosets in (2.8).

However, if N is a normal subgroup of G, then right and left cosets of N do agree:

PROPOSITION 2.9. *A subgroup H of a group G is normal if and only if $aH = Ha$ for each $a \in G$.*

PROOF. Suppose that H is a normal subgroup of G. Then for any $h \in H$ and any $a \in G$, we have

$$ah = (aha^{-1})a,$$

where $k = aha^{-1}$, the conjugate of h, becomes an element of H because H is normal. This shows that $aH \subseteq Ha$. Similarly, $Ha \subseteq aH$ holds, and this proves the $\Longrightarrow$ direction. Conversely, suppose that $aH = Ha$ for each $a \in G$. Let $a \in G$, $h \in H$ be any elements. Then since $aH = Ha$, $ah = h'a$; hence $aha^{-1} = h' \in H$, and this proves the $\Longleftarrow$ direction. □

COROLLARY 2.12. *Every subgroup H of a group G of index 2 in G is normal.*

PROOF. H has exactly two left cosets $1H$ and gH. Then we easily see that $1H = H1$ and $gH = G - H = Hg$ holds. □

COROLLARY 2.13. *Let H be a subgroup of a group G. If H is not normal, then there exists a non-identity element $a \in G$ such that $aH \not\subseteq Ha$ and moreover $aH \not\subseteq Hb$ for any $b \in G$ which is not right congruent to a.*

PROOF. Suppose that H is not normal. Then there exist $h \in H$ and a non-identity element $a \in G$ such that

$$k = aha^{-1}$$

is not an element of H. Then since $ah = (aha^{-1})a = ka$, we see that ah is in aH but it is not in Ha, for, if $ah = ka$ were an element of Ha, then $ka = h'a$ for some $h' \in H$; hence $k = h' \in H$, which contradicts our hypothesis that k is not an element of H, and this shows that $aH \not\subseteq Ha$. On the other hand, we note that aH and Ha do have a common element, i.e., the element a. Thus, by (2.7), we see that aH cannot be a subset of some other distinct right coset Hb of G for any b which is not right congruent to a, i.e., for any b such that $b \neq ha$ for any $h \in H$. This proves the assertion. □

An equivalent version of this corollary is the following:

COROLLARY 2.14. *If H is not normal, then the partition of G into left cosets is not the same as the partition of G into right cosets.* □

EXAMPLE 2.38. *Equations in (2.8) and (2.10) tell us that the subgroup $H = \{1, xy\}$ is not a normal subgroup of S_3.*

To have a better understanding on the concept of normality of a subgroup H of a group G, we close this section by proving the following:

PROPOSITION 2.10. *Let H be a subgroup of a group G. Then the following statements are equivalent:*

(1) *Left and right congruence relation with respect to H coincide. Namely, they define the same partition of G.*
(2) *Every left coset aH is equal to some right coset Hb. Namely, each aH is equal to some Hb.*
(3) *$aH = Ha$ for each $a \in G$.*
(4) *For each $a \in G$, $aHa^{-1} \subseteq H$.*
(5) *For each $a \in G$, $aHa^{-1} = H$.*

PROOF. Part (1) $\Longrightarrow$ Part(2) is trivial.

Part (2) $\Longrightarrow$ Part(3): If $aH = Hb$ for some $b \in G$, then a is a common element of both Ha and Hb; hence $Hb = Ha$; consequently $aH = Ha$.

Part (3) $\Longrightarrow$ Part(1) is trivial.

Part (3) $\Longrightarrow$ Part (4) is trivial.

Part (4) $\Longrightarrow$ Part (5). We have $aHa^{-1} \subseteq H$. Then since Part (4) also holds for a^{-1}, we also have $a^{-1}Ha \subseteq H$. For the reverse inclusion, let h be any element of H. Then

$$h = a(a^{-1}ha)a^{-1} \in aHa^{-1};$$

hence $H \subseteq aHa^{-1}$.

Part (5) $\Longrightarrow$ Part (3) is trivial, because every $ah = h'a$ for some $h' \in H$ by Part (5). □

REMARK 2.28. *One more equivalent statement for H being normal in G is given in Proposition 2.18.*

Therefore,

DEFINITION 2.28. *A subgroup H of a group G is* **normal** *if it satisfies one of the equivalent statements in Proposition 2.10.*

COROLLARY 2.15. *Let N be the kernel of a group homomorphism $\varphi : G \longrightarrow H$. Then $aN = \varphi^{-1}(\varphi(a)) = Na$, for any $a \in G$.*

PROOF. We have, for instance: $x \in \varphi^{-1}(\varphi(a))$ if and only if $\varphi(x) = \varphi(a)$ if and only if $xa^{-1} \in N$ if and only if $x \in Na$. □

Exercises for Chapter 2.7

EXERCISE 2.55. *Show that the number of left cosets of a subgroup H of a group G of finite order is the same as that of right cosets of H.*

EXERCISE 2.56. *Prove that any group of order 6 is isomorphic to either $\mathbb{Z}_6$ or S_3. (Hint: G contains (i) an element of order 6 , (ii) an element of order 3 but none of order 6, or (iii) all elements of G have order 1 or 2.*

EXERCISE 2.57. *Prove that any group of order 10 is isomorphic to $\mathbb{Z}_{10}$ or D_5, where D_n is the dihedral group of order $2n$ (Definition 3.6)*

EXERCISE 2.58. *Determine all groups of order $2p$, where p is a prime number.*

EXERCISE 2.59. *Show that A_3 is the only proper normal subgroup of S_3.*

EXERCISE 2.60. *Find an example of a subgroup of index 3 which is not normal.*

EXERCISE 2.61. *Prove that the cyclic group of order p is the only proper normal subgroup of D_p, where p is an odd prime number. (Definition 3.6)*

EXERCISE 2.62. *List all normal subgroups of D_4. (Definition 3.6)*

EXERCISE 2.63. *List all normal subgroups of D_6. (Definition 3.6)*

EXERCISE 2.64. *Let S be a subgroup of a group G. Prove that the intersection of all conjugates of S is a normal subgroup of G.*

8. Restriction of a Homomorphism to a Subgroup

Suppose that you are given a group of a large order, including an infinite order. If it is a finitely generated abelian group, then we may answer its algebraic structure, applying Linear Algebra, called the Theory of Modules. If we do not know whether it is abelian or not, it is usually quite difficult even when it is a finite group. The usual way to have a better understanding of a complicated group is to study some less complicated subgroups, and this approach may be one of the best approaches in studying a complicated group. Although there is no general rule for obtaining information for a given group from an understanding of its subgroups, this approach is sometimes quite fruitful.

But whenever a new construction with groups is made, we should study its effect on subgroups, and this is all about *restriction to a subgroup.* We start our discussion with a simple situation.

PROPOSITION 2.11. *The intersection $H \cap N$ of a subgroup H of a group G with a normal subgroup N of G is a normal subgroup of H.*

PROOF. We need to show that the subgroup $H \cap N$ of H by Lemma 2.6 is normal in H. For this, let $a \in H$ and $k \in H \cap N$ be any elements, respectively. Then since N is normal in G and $k \in H$, $aka^{-1} \in H \cap N$, and hence by Part (4) of Proposition 2.10 we are done. □

REMARK 2.29. *Before we generalize this result to a more abstract version under a group homomorphism, we note that the counting formula in (2.9) and Lagrange's Theorem (Corollary 2.8) are usefully applied to get information about the intersection, provided that G is a finite group. That is, since $H \cap N$ is a subgroup of both H and N, its order $|H \cap N|$ divides both of the orders $|H|$ and $|N|$. Thus, if* $\gcd(|H|, |N|) = 1$, *then we can conclude that $H \cap N = \{e\}$.*

In terms of a group homomorphism, Proposition 2.11 says the following: Let H and N be two subgroups of a group G. Then we have an obvious group homomorphism:

$$i : H \hookrightarrow G,$$

and $H \cap N$ is nothing but the inverse image $i^{-1}(N)$ of N in H. Then Lemma 2.6 says that $i^{-1}(N)$ is a subgroup of H and moreover, Proposition 2.11 says that if N is normal in the target G, then its inverse image, $i^{-1}(N)$, is normal in the domain H.

DEFINITION 2.29. *Let $\varphi : G \longrightarrow G'$ be a group homomorphism, H a subgroup of G. Then the composition:*

$$\varphi \circ i : H \hookrightarrow G \longrightarrow G'$$

*is called the **restriction** of φ to H and is denoted by*

$$\varphi|_H : H \longrightarrow G'.$$

This simply means that we take the same map φ but *restrict* its domain to H. That is, $\varphi|_H(h) = \varphi(h)$ for each $h \in H$. The restriction is obviously a group homomorphism because it is a composition of two such homomorphisms, and its kernel is the intersection of $\ker \varphi$ with H:

$$\ker \varphi|_H = (\ker \varphi) \cap H. \tag{2.11}$$

That is, $\varphi(h) = 1$ for any $h \in H$ if and only if $h \in \ker(\varphi)$. This is because

$$\ker \varphi|_H = (\varphi \circ i)^{-1}(e) = i^{-1}(\varphi^{-1}(e)) = i^{-1}(\ker \varphi) = (\ker \varphi) \cap H.$$

REMARK 2.30. *We note that the counting formula again helps us to derive a little bit more out of the restriction map when G and G' are finite groups: Since the image of $\varphi|_H$ is $\varphi(H)$, we know by Corollary 2.11 that the order $|\varphi(H)|$ should divide both $|H|$ and $|G'|$. Thus, if $\gcd(|H|, |G'|) = 1$, then we can conclude that $|\varphi(H)| = 1$; hence H must be contained in the kernel, $\ker \varphi$.*

EXAMPLE 2.39. *We have seen in Example 2.21 that the sign of a permutation is a group homomorphism:*

$$sign : S_n \longrightarrow \{\pm 1\},$$

whose kernel is the alternating group A_n. Thus, A_n is a normal subgroup of S_n.

This allows us to make some conclusion about permutations of odd order: Since the group $\{\pm 1\}$ has order 2, if a subgroup H of S_n has odd order, then the restriction of this homomorphism to H has the trivial kernel, by Remark 2.30 and Corollary 2.11; hence H is contained in A_n. Thus, H must consist of even permutations only. In particular, if H is the cyclic subgroup generated by a permutation p whose order in the group is odd, H must consists of even permutations. It follows that every permutation of odd order is an even permutation.

On the other hand, we can not draw any conclusion from the counting formula about permutations of even order. They may be odd or even.

Now, we are going to generalize Proposition 2.11. Let $\varphi : G \longrightarrow G'$ be a group homomorphism and H' a subgroup of G'. In contrast to the previous case, we want to "restrict" φ to H' in such a way that we can get a natural group homomorphism to H'. The best way for this is to consider the inverse image $\varphi^{-1}(H')$ of H'.

$$\varphi^{-1}(H') = \{g \in G : \varphi(g) \in H'\}.$$

It is a subset of the domain G. In fact, we have a lot more than this:

PROPOSITION 2.12. *Let $\varphi : G \longrightarrow G'$ be a group homomorphism and H' a subgroup of G'. Then*

(1) *$\varphi^{-1}(H')$ is a subgroup of G.*
(2) *If H' is normal in G', then so is $\varphi^{-1}(H')$ in G.*
(3) $\ker \varphi \subseteq \varphi^{-1}(H')$.
(4) *The restriction of φ to $\varphi^{-1}(H')$ defines a group homomorphism*

$$\varphi^{-1}(H') \longrightarrow H',$$

whose kernel is $\ker\varphi$.

PROOF. Part (1). The identity element 1 of G is in $\varphi^{-1}(H')$ because $\varphi(1) = 1 \in H'$. If $u, v \in \varphi^{-1}(H')$, then $g(u) \in H'$ and $g(v) \in H'$; hence $g(u)g(v) = g(uv) \in H'$; hence their product $uv \in \varphi^{-1}(H')$. If $u \in \varphi^{-1}(H')$, then $\varphi(u) \in H'$; hence $\varphi(u)^{-1} = \varphi(u^{-1}) \in H'$; hence its inverse $u^{-1} \in \varphi^{-1}(H')$. Thus, by Definition 2.7 $\varphi^{-1}(H')$ is indeed a subgroup of G.

Part (2). Suppose furthermore that H' is normal in G'. Then for any $n \in \varphi^{-1}(H')$ and any $g \in G$, we need to show that $gng^{-1} \in \varphi^{-1}(H')$. But, we have

$$\varphi(gng^{-1}) = \varphi(g)\varphi(n)\varphi(g)^{-1} \in H',$$

because $\varphi(n) \in H'$. Thus, $gng^{-1} \in \varphi^{-1}(H')$, and by Part (4) of Proposition 2.10, we conclude that $\varphi^{-1}(H')$ is normal in G.

Part (3). If $u \in \ker\varphi$, then $\varphi(u) = 1 \in H'$; hence $u \in \varphi^{-1}(H')$.

Part (4) must be clear by the same argument as in (2.11), this time with $H = \varphi^{-1}(H')$. □

EXAMPLE 2.40. *The set of positive real numbers is a normal subgroup of* $\mathbb{R}^\times$*; hence its inverse image under the group homomorphism*

$$\det : GL_n(\mathbb{R}) \longrightarrow \mathbb{R}^\times,$$

consisting of those invertible $n \times n$ *matrices with positive determinant, is a normal subgroup of* $GL_n(\mathbb{R})$.

PROPOSITION 2.13. *(Correspondence Theorem) Let* $\varphi : G \longrightarrow G'$ *be a* surjective *homomorphism of groups with kernel* N. *Then the assignment*

$$H \longmapsto \varphi(H) \qquad \text{and} \qquad H' \longmapsto \varphi^{-1}(H')$$

is a one-to-one correspondence between the set of subgroups H' *of* G' *and the set of subgroups of* G *which contain* N. *Moreover, normal subgroups of* G *containing* N *correspond to normal subgroups of* G'.

PROOF. In general, $H \subseteq \varphi^{-1}(\varphi(H))$ and $\varphi(\varphi^{-1}(H')) \subseteq H'$ hold. Then since φ is surjective, $H' \subseteq \varphi(\varphi^{-1}(H'))$ holds. On the other hand, since $N \subseteq H$, $\varphi^{-1}(\varphi(H)) \subseteq H$ holds. This is because: if $x \in \varphi^{-1}(\varphi(H))$ then $\varphi(x) = \varphi(h)$ for some h in H; hence $xh^{-1} \in \ker\varphi = N \subseteq H$; hence $x \in H$. The proof of the last statement is left to the readers as an exercise. □

Exercises for Chapter 2.8

EXERCISE 2.65. *Let G and G' be finite groups whose orders are relatively prime. Prove that the only group homomorphism from G to G' is the trivial one, sending everything to* 1.

EXERCISE 2.66. *Show that a group of order* 30 *can have at most* 7 *subgroups of order* 5.

EXERCISE 2.67. *Complete the proof of the Correspondence Theorem, Proposition 2.13.*

9. Products of Groups

In this section, we are going to define a group structure on the Cartesian product of two groups G and G'. This concept sometimes enable us to describe the structure of certain finite groups in detail.

DEFINITION 2.30. *Let G and G' be two groups. We define multiplication of pairs (g, g') and (h, h') in $G \times G'$ by the rule:*

$$(g, g'), (h, h') \rightsquigarrow (gh, g'h').$$

LEMMA 2.28. *Under this definition, $G \times G'$ is a group of order $|G| \cdot |G'|$, called the* **product** *of G and G'.*

PROOF. $(1, 1)$ is the identity element of $G \times G'$, the inverse of (g, h) is (g^{-1}, h^{-1}), and the associativity follows from the ones in G and G'. Its order is obviously the product of the orders of G and G'. □

We then note that there are four natural group homomorphism:

$$\begin{array}{ccccccc} G & \overset{i}{\hookrightarrow} & G & \times & G' & \overset{i'}{\hookleftarrow} & G' \\ & & \swarrow_{p} & & \searrow^{p'} & & \\ & G & & & & G' & \end{array},$$

defined by the rule

$$\begin{aligned} i(g) &= (g, 1), \\ i'(g') &= (1, g'), \\ p(g, g') &= g, \\ p'(g, g') &= g'. \end{aligned}$$

Then, the maps i and i' are injective so that we may identify G, G' with the subgroup $G \times 1$ and $1 \times G'$ of $G \times G'$, respectively. The maps p and p', called **projections**, are surjective with kernels $\ker p = 1 \times G'$, $\ker p' = G \times 1$, respectively. Thus, the subgroups $G \times 1$ and $1 \times G'$ are always normal in $G \times G'$. In other words, we have constructed a new group $G \times G'$ such that G and G' are normal in $G \times G'$ under the identifications through i and i'.

We then have:

PROPOSITION 2.14. *(The mapping property of products) Group homomorphisms $\Phi : H \longrightarrow G \times G'$ are in one-to-one correspondence with pairs (φ, φ') of group homomorphisms*

$$\varphi : H \longrightarrow G, \qquad \varphi' : H \longrightarrow G',$$

and $\ker \Phi = (\ker \varphi) \cap (\ker \varphi')$.

PROOF. Given Φ, define φ, φ' respectively by composition with the projections as follows:

$$\varphi = p \circ \Phi, \qquad \varphi' = p' \circ \Phi.$$

Then, if we write $\Phi(h) = (x, y)$, then $p \circ \Phi(h) = x$ and $p' \circ \Phi(h) = y$ so that we have

$$\Phi = (p \circ \Phi, p' \circ \Phi).$$

Conversely, given a pair (φ, φ') of group homomorphisms, define Φ by

$$\Phi(h) := (\varphi(h), \varphi'(h)).$$

Then Φ is clearly a group homomorphism. One easily checks that the composition of these two assignments in both directions is the identity map. Moreover, $\Phi(h) = (1, 1)$ if and only if $\varphi(h) = 1$ and $\varphi'(h) = 1$, showing that $\ker \Phi = (\ker \varphi) \cap (\ker \varphi')$. □

We note that if we are able to express a given group G as a product group, $G \simeq H \times H'$, for some groups H and H', then since the order of H and H' are expected to be smaller than that of G, their structures are going to be simpler than that of G, where the relation between $H \times H'$ and its factors H, H' is already well-understood, so that we can have a better understanding of the group G itself. In reality, it is quite rare that a given group is a product group, but occasionally it does happen, and here is such a case:

If C_{rs} is a cyclic group of order rs, generated by x, then $\{x^{ri} : i = 1, ..., s\}$ is a cyclic subgroup C_s of order s; hence we have a group homomorphism $\varphi : C_{rs} \longrightarrow C_s$ defined by sending

$$x^i \longmapsto x^{ri}.$$

Then φ is clearly surjective with kernel, $C_r = \{x^{si} : i = 1, ..., r\}$. Under this setting, we have:

PROPOSITION 2.15. *Let r, s be two natural numbers such that*

$$\gcd(r, s) = 1.$$

Then a cyclic group C_{rs} of order rs is isomorphic to the product of cyclic groups C_r of order r and C_s of order s,

$$C_{rs} \simeq C_r \times C_s, \qquad \gcd(r, s) = 1.$$

PROOF. Write $C_{rs} = \{1, x, x^2, ..., x^{rs-1}\}, C_r = \{1, y = x^s, ..., y^{r-1}\}, C_s = \{1, z = x^r, ..., z^{s-1}\}$, and we define, using Proposition 2.14, a map φ :

$C_{rs} \longrightarrow C_r \times C_s$ by sending

$$x^i \longmapsto (y^i, z^i).$$

Then φ is clearly a group homomorphism with $\ker\varphi = \{x^i \in C_{rs} : y^i = 1$ and $z^i = 1\}$. We need to figure out which of the rs elements of C_{rs} are in $\ker(\varphi)$. For this, we note that $y^i = 1$ if and only if i is divisible by r, and similarly $z^i = 1$ if and only if i is divisible by s. Thus, $\operatorname{lcm}(r, s)$ divides i. Then since $\gcd(r, s) = 1$, $\operatorname{lcm}(r, s) = rs$. Thus, none of the numbers between 1 and $rs - 1$ can be divisible by both r and s; hence $1 = x^{rs}$ is the only element of C_{rs} of which image under φ is 1. That is, $\ker\varphi = \{1\}$. This proves that the map φ is injective. Then since they have the same order, the injective map φ is in fact a surjective map; hence it is an isomorphism. □

EXAMPLE 2.41. *A cyclic group C_6 is isomorphic to the product group, $C_2 \times C_3$, because* $6 = 2 \cdot 3$ *with* $\gcd(2, 3) = 1$.

On the other hand, we have:

EXAMPLE 2.42. *A cyclic group C_4 is not isomorphic to $C_2 \times C_2$, because one can easily verify that C_4 contains two elements of order 4 and every element of $C_2 \times C_2$ has order either 1 or 2.*

Thus the requirement,

$$\gcd(r, s) = 1,$$

is essential in Proposition 2.15.

In order to characterize the product groups in terms of normal subgroups, we need to introduce a definition:

DEFINITION 2.31. *Let A and B be subsets of a group G. Define*

$$AB = \{g \in G : g = ab \textit{ for some } a \in A, b \in B\}.$$

We note that the subset AB of G is not a group in general, even if A and B are subgroups of G.

The following proposition characterizes product groups:

PROPOSITION 2.16. *Let H and K be subgroups of a group G.*

(1) *If $H \cap K = \{1\}$, then the product map $p : H \times K \longrightarrow G$ defined by sending*

$$(h, k) \longmapsto hk$$

is injective, and its image is the subset HK.

(2) *If either H or K is a* normal *subgroup of G, then $HK = KH$, and HK is a subgroup of G.*

(3) *If*
   (a) *both H and K are normal in G,*
   (b) *$H \cap K = \{1\}$, and*
   (c) *$HK = G$,*

then G is isomorphic to the product group $H \times K$.

REMARK 2.31. *We should note that the product map p in Part (1) is not a group homomorphism in general, unless $hk = kh$ for each $h \in H$,each $k \in K$. To make it happen we need the normality of both H and K in G, as we shall see it in the proof below in detail.*

PROOF. Part (1). Let (h_1, k_1) and (h_2, k_2) be elements of $H \times K$ such that $h_1k_1 = h_2k_2$. Then $h_2^{-1}h_1 = k_2k_1^{-1} \in H \cap K = \{1\}$; hence $h_1 = h_2$ and $k_1 = k_2$. Thus, the map p is injective. Its image is obviously the set HK.

Part (2). We suppose that H is normal in G. Then for any $h \in H$ and any $k \in K$, $khk^{-1} \in H$; hence $khk^{-1} = h'$; hence $kh = h'k$, and this shows that $KH \subseteq HK$. Changing k with k^{-1}in this argument, we easily conclude that $HK \subseteq KH$.

Now, we note that $1 \in HK$. For any hk and $h'k'$, since $HK = KH$ we have

$$hk \cdot h'k' = h(kh')k' = h(h_1k_1)k' \in HK,$$

so HK is closed under multiplication induced from the one in G. Now, for any $hk \in HK$, $(hk)^{-1} = k^{-1}h^{-1} = h_1k_1 \in HK$; hence $(hk)^{-1} \in HK$. This proves that HK is a subgroup of G. The proof is similar when K is normal.

Part (3). First we are going to prove that

$$hkh^{-1}k^{-1} = 1, i.e., hk = kh,$$

for each $h \in H$, each $k \in K$ in order to make the product map p a group homomorphism. Since both H and K are normal in G, we see that

$$K \ni (hkh^{-1})k^{-1} = h(kh^{-1}k^{-1}) \in H.$$

Then since $H \cap K = \{1\}$, we conclude that $hkh^{-1}k^{-1} = 1$. Thus, the product map p sending $(h, k) \longmapsto hk$ is indeed a group homomorphism under the two assumptions (a) and (b):

$$\begin{aligned} p(h_1, k_1)p(h_2, k_2) &= (h_1k_1) \cdot (h_2k_2) \\ &= h_1(k_1h_2)k_2 = h_1(h_2k_1)k_2 \\ &= (h_1h_2) \cdot (k_1k_2) = p(h_1h_2, k_1k_2) \\ &= p((h_1, k_1)(h_2, k_2)). \end{aligned}$$

Then we have already seen in Part (1) that p is injective, and by assumption (c), p is surjective. Thus, p is an isomorphism. □

REMARK 2.32. *Proposition 2.16 is, for instance, utilized in the proof of Corollary 4.5.*

Exercises for Chapter 2.9

EXERCISE 2.68. *Consider a product group $G \times G$.*

(1) *Show that the set D of all diagonal pairs (g, g) of $G \times G$ is a subgroup of $G \times G$.*
(2) *Prove that G is abelian if and only if D is normal in $G \times G$.*

EXERCISE 2.69. *Prove that the center of the product group is the product of their centers.*

EXERCISE 2.70. *Prove that the product HK of two subgroups H, K of a group G is a subgroup of G if and only if $HK = KH$.*

EXERCISE 2.71. *Let G be a group containing normal subgroups of order* 3 *and* 5, *respectively. Prove that it contains an element of order* 15.

EXERCISE 2.72. *Let G be a finite group whose order n is a product of two numbers, $n = ab$. Suppose that G has subgroups H, K of order a and b, respectively, such that $H \cap K = \{1\}$. Prove that $G = HK$.*

EXERCISE 2.73. *Prove that $G \times \{1'\}$ is a normal subgroup of $G \times G'$.*

10. Modular Arithmetic

REMARK 2.33. *We first note that it was Gauss (1777 - 1855) who first introduced such a notation $\equiv$ in his very famous book,* Disquisitiones Arithmeticae, *published in 1801, where he treated congruence properties of integers in a greatly systematic manner. Ever since then, the definition of congruence of integers remained as one of the most important concepts in Number Theory.*

In this section we fix a natural number n, unless otherwise mentioned.

DEFINITION 2.32. *Two integers a, b are said to be* **congruent modulo** *n, written by*

$$a \equiv b (\operatorname{mod} n),$$

if n divides the difference $a - b$, that is, $b = a + kn$ for some integer k.

In terms of modern terminology, we can easily verify that

LEMMA 2.29. *The congruence relation is an equivalence relation on the set $\mathbb{Z}$ of integers.* $\square$

Thus, we may consider the equivalence classes, called **congruence classes modulo** n or **residue classes modulo** n, defined by this relation, which we have already studied in Section 6 of this Chapter.

NOTATION 2.16. *The congruence class of an integer a is denoted by $\bar{a}$,*

$$\bar{a} = \{..., a - 2n, a - n, a, a + n, a + 2n, ...\}.$$

Thus, any element $b \in \bar{a}$ gives $\bar{b} = \bar{a}$, and conversely, $\bar{b} = \bar{a}$ means that $b - a$ is divisible by n. In particular, the congruence of 0 is the subgroup

$$\bar{0} = n\mathbb{Z}$$

of the additive group $\mathbb{Z}^+$, which consists of all multiples of n, and other congruence classes $\bar{a}$ are the cosets of this subgroup:

$$\bar{a} = a + n\mathbb{Z}.$$

Then the following proposition says that the n integers $0, 1, ..., n-1$ are natural representatives for the congruence classes.

PROPOSITION 2.17. *There are n congruence classes modulo n:*

$$\bar{0}, \bar{1}, ..., \overline{n-1}.$$

PROOF. Given an integer a, we apply the division with remainder,

$$a = nq + r,$$

for the unique quotient q and the unique remainder r with

$$0 \leq r \leq n - 1.$$

Then since $a - r = nq$, a is congruent to the remainder r. Thus, $\bar{a} = \bar{r}$. On the other hand, if a and b are distinct integers strictly less than n, then their difference $a - b$ is also less than n; hence not divisible by n. Thus, $a \not\equiv b$ modulo n, and this means that $\bar{a} \neq \bar{b}$. This proves that the n classes $\bar{0}, \bar{1}, ..., \overline{n-1}$ are all distinct. □

COROLLARY 2.16. *The index $[\mathbb{Z} : n\mathbb{Z}]$ of the subgroup $n\mathbb{Z}$ in $\mathbb{Z}$ is n.* □

In what follows, we are going to define addition and multiplication of congruence classes in terms of congruence modulo n so that the set of congruence classes becomes a commutative ring. In other words, the usual addition and multiplication of integers do allow us to define in a natural manner addition and multiplication of congruence classes, and because of this reason, the new operations are usually called to be the induced binary operations from those two on $\mathbb{Z}$. This process is very important because it in fact shows everything on how we should define addition and multiplication of equivalence classes on a group or a commutative ring in general, and this matter is fully discussed in the next section.

DEFINITION 2.33. *Let $\bar{a}$ and $\bar{b}$ be two congruence classes of integers. We define their* ***sum*** *to be the congruence class of their sum $a+b$ in $\mathbb{Z}$, and their* ***product*** *is defined to be the congruence class of their product ab in $\mathbb{Z}$. That is,*

$$\bar{a}+\bar{b} := \overline{a+b}, \qquad \bar{a}\bar{b} := \overline{ab}. \tag{2.12}$$

Then since $\bar{a}$ and $\bar{b}$ usually can have different representatives, say a' and b', we need to verify that these two new binary operations are **well-defined**:

LEMMA 2.30. *If $a \equiv a' \bmod n$ and $b \equiv b' \bmod n$, then $a+b \equiv a' + b' \bmod n$ and $ab \equiv a'b' \bmod n$.*

PROOF. By hypothesis, $a' = a + nr$ and $b' = b + ns$ for some integers r, s; hence $a' + b' = a + b + n(r+s)$, showing that $a + b \equiv a' + b' \bmod n$. Moreover, $a'b' = (a+nr)(b+ns) = ab + n(as + rb + nrs)$, showing that $ab \equiv a'b' \bmod n$. □

LEMMA 2.31. *The associative, distributive, commutative laws hold for the laws of compositions in (2.12).*

PROOF. They hold because they hold for addition and multiplication of integers. For the distributive law, we see that

$$\bar{a}(\bar{b}+\bar{c}) = \overline{a(b+c)} = \overline{ab+ac} = \overline{ab} + \overline{ac} = \bar{a}\bar{b} + \bar{a}\bar{c}.$$

Similarly, one can easily verify other laws. □

NOTATION 2.17. *The set of congruence classes modulo n is usually denoted by*

$$\mathbb{Z}/n\mathbb{Z}.$$

Then the above lemma equivalently says that $\mathbb{Z}/n\mathbb{Z}$ is a commutative ring together with a canonical ring homomorphism to it from $\mathbb{Z}$,

$$\mathbb{Z} \longrightarrow \mathbb{Z}/n\mathbb{Z},$$

which sends $a \longmapsto \bar{a} = a + n\mathbb{Z}$. We note that $\bar{0} = 0 + n\mathbb{Z}$ is the additive identity and $\bar{1} = 1 + n\mathbb{Z}$ is the multiplicative identity of the ring $\mathbb{Z}/n\mathbb{Z}$. The ring $\mathbb{Z}/n\mathbb{Z}$ is called a *quotient ring*.

REMARK 2.34. *The reader should note that the ring homomorphism:* $\mathbb{Z} \longrightarrow \mathbb{Z}/n\mathbb{Z}$ *defined above is a main motivation for the definition of a ring homomorphism between two rings.*

Exercises for Chapter 2.10

EXERCISE 2.74. *Construct addition and multiplication table for* $\mathbb{Z}_6$ *and show that multiplication in* $\mathbb{Z}_6$ *does not satisfy the cancellation law: if* $ab = ac$ *and* $a \neq 0$, *then* $b = c$.

EXERCISE 2.75. *Prove that* $n^2 \equiv 0, 1,$ *or* $4 \pmod 8$.

EXERCISE 2.76. *Prove that the square* a^2 *of an integer* a *is congruent to either* 0 *or* 1 *modulo* 4.

EXERCISE 2.77. *Prove that no number* $k \equiv 7 \pmod 8$ *can be expressed as the sum of three squares.*

EXERCISE 2.78. *Prove that the sum of the cubes of any three consecutive integers is divisible by* 9.

EXERCISE 2.79. *Prove that every integer* a *is congruent to the sum of its decimal digits modulo* 9.

EXERCISE 2.80. *In the usual decimal notation, the addition and multiplication of large numbers may be checked by the rule of "**casting out 9's**": Replace each number to be added by the sum of its digits, add these, and compare with the sum of the digits of the proposed answer (in comparing, use the sum of the digits of the sum of the digits...). Explain why this rule works.*

11. Quotient Groups

We have seen in the last section that the congruence classes of integers modulo n are the cosets of the subgroup $n\mathbb{Z}$ of $\mathbb{Z}$, where addition of congruence classes, which is induced from the law of addition of two integers, is a law of composition on the set of these cosets. In this section, we are going to learn that a law of composition can be defined on the cosets of a normal subgroup N of a group G, which makes the set of cosets into a group, called a **quotient group**.

EXAMPLE 2.43. *We know that an angle in the plane has period* 2π, *so that two angles are the same if and only if their difference is divisible by* 2π. *Thus, we have a group homomorphism* $\mathbb{R}^+ \longrightarrow \mathbb{R}^+/2\pi\mathbb{Z}$ *which sends a positive real number*

$$a \longmapsto \theta,$$

where θ *is the unique remainder in* $a = 2\pi q + \theta$ *with* $0 \leq \theta < 2\pi$. *Here, we again note that, since* $\mathbb{R}^+$ *is an abelian group, any subgroup of* $\mathbb{R}^+$ *is automatically normal in* $\mathbb{R}^+$.

LEMMA 2.32. *Let* N *be a normal subgroup of a group* G. *Then the product of two cosets* aN, bN *of* N *is again a coset,*

$$aNbN = abN,$$

and this coset multiplication is well-defined.

PROOF. We first note that $NN = N$ since N is a subgroup of G. Now, let $an_1 \in aN, bn_2 \in bN$ be any elements of the cosets. Then since N is normal in G, $n_1 b = bn_1'$ for some $n_1' \in N$ (Proposition 2.10). Thus, we have

$$an_1 \cdot bn_2 = a(n_1 b)n_2 = a(bn_1')n_2 = ab(n_1' n_2).$$

Then since N is a subgroup $NN = N$, these equalities exactly mean that $aNbN = abN$. For instance, $abn = abn1 = an'b1$ for some $n' \in N$, so that $abN \subseteq aNbN$.

Now we need to prove that the formula is well-defined. For this, we note that if $aN = \alpha N$ and $bN = \beta N$, then since $xN = Nx$ by Proposition

2.10 we have:

$$abn = an_1b = \alpha n_2b = \alpha bn_3 = \alpha\beta n_4,$$

for some $n, n_1, n_2, n_3, n_4 \in N$, so that $abN \subseteq \alpha\beta N$ holds. Similarly, $\alpha\beta N \subseteq abN$ holds. Thus, we have:

$$aNbN = abN = \alpha\beta N = \alpha N\beta N.$$

This proves that the formula in the lemma is indeed well-defined. □

In the above lemma, the assumption that N is normal in G is absolutely needed: If N is not a normal subgroup of G, then there are elements $n \in N$ and $g \in G$ such that $gng^{-1} \notin N$. Then consider the product of two cosets,

$$gNg^{-1}N.$$

We obviously have: $g1g^{-1}1 = 1 \in N$. Thus, if $gNg^{-1}N$ is contained in any coset of N, it must be $N = 1N$. But, this is impossible because $gng^{-1}1$ is not an element of N. This proves that if N is not a normal subgroup of G then the product of cosets $gNg^{-1}N$ is not contained in any coset of N. In fact, we have more than this:

PROPOSITION 2.18. *Let N be a subgroup of a group G. Then the coset multiplication is* well-defined *by the formula*

$$aNbN = abN$$

if and only if N is a normal *subgroup of G.*

REMARK 2.35. *One more equivalent statement for N being normal in G is given in Corollary 4.6.*

PROOF. The $\Longleftarrow$ direction is already established by Lemma 2.32, so we need to prove the $\Longrightarrow$ direction. For any $x \in aN$, we have $xN = aN$, and $xNa^{-1}N = xa^{-1}N$. Moreover, we have $aNa^{-1}N = aa^{-1}N = N$. Then since the formula is well-defined, we must have $xa^{-1}N = N$; hence $xa^{-1} = n$ for some $n \in N$; hence $x \in Na$. This proves that $aN \subseteq Na$. For the reverse inclusion, let $x \in Na$. Then $x = na$ for some $n \in N$; hence $x^{-1} = a^{-1}n^{-1} \in a^{-1}N$; hence $x^{-1}N = a^{-1}N$. Then since the formula is well-defined, $x^{-1}aN = x^{-1}NaN = a^{-1}NaN = N$; hence $x^{-1}a = n_1$ for some $n_1 \in N$; hence $x = an_1^{-1} \in aN$. This prove that $Na \subseteq aN$. Thus, by Proposition 2.10 we are done. $\square$

NOTATION 2.18. *As in the case of congruence classes modulo n where the set of cosets of $n\mathbb{Z}$ is denoted by $\mathbb{Z}/n\mathbb{Z}$, the set of cosets of a normal subgroup N of a group G is denoted by*

$$G/N.$$

Then we have a natural surjective map $\pi : G \longrightarrow G/N$ sending each $a \in G$

$$a \longmapsto \bar{a} = aN. \tag{2.13}$$

LEMMA 2.33. *Let G be a group, S any set with a law of composition. Suppose that we have a surjective map $\varphi : G \longrightarrow S$ such that $\varphi(ab) = \varphi(a)\varphi(b)$ for each $a, b \in G$. Then S is a group.*

PROOF. For any $s \in S$, $s = \varphi(s')$ for some $s' \in G$. Thus, $\varphi(1)$ is the identity for the law of composition because $s = \varphi(s') = \varphi(1s') = \varphi(1)\varphi(s') = \varphi(1)s$. Now, let $s, t, u \in S$. Then $s = \varphi(s'), t = \varphi(t'), u = \varphi(u')$ for some $s', t', u' \in G$. Thus, the associative law holds:

$$\begin{aligned} s(tu) &= \varphi(s')(\varphi(t')\varphi(u')) = \varphi(s')(\varphi(t'u')) = \varphi(s't'u') \\ &= \varphi(s't')\varphi(u') = (\varphi(s')\varphi(t'))\varphi(u') = (st)u. \end{aligned}$$

Finally, for any $s \in S$, $s = \varphi(s')$ for some $s' \in G$. Thus, $\varphi(s'^{-1})s = \varphi(s'^{-1})\varphi(s') = \varphi(1)$; hence $\varphi(s'^{-1})$ is the inverse of s. This proves that S is a group. $\square$

THEOREM 2.1. *Let N be a normal subgroup of a group G. Then G/N is a group under the law of composition defined in Lemma 2.32, and the surjective map π in (2.13) is a group homomorphism with kernel N.*

PROOF. The surjective map $\pi : G \longrightarrow G/N$ is clearly a group homomorphism: $\pi(ab) = abN = aNbN = \pi(a)\pi(b)$, and those $a \in G$ such that $\pi(a) = \bar{1} = 1N = N$ must only belong to N, because we have a partition on G by the cosets of N. Finally, G/N is a group by Lemma 2.33. $\square$

COROLLARY 2.17. *The order of G/N is the index $[G : N]$ of N in G.*

PROOF. By Corollary 2.11, we have this result. □

COROLLARY 2.18. *Every normal subgroup of a group is the kernel of a group homomorphism.* □

EXAMPLE 2.44. *Let $G = \mathbb{R}^\times$ be the multiplicative group of non-zero real numbers, P the subgroup of positive real numbers. Since G is an abelian group, P is automatically a normal subgroup of G, and hence G/P is a group consisting of two cosets, namely P and $-P = \{$negative real numbers$\}$; hence $G/P \simeq \{1, -1\}$, a cyclic group of order 2 generated by -1: $(-1)^2 = 1$.*

We recall that the kernel of a group homomorphism is always a normal subgroup of the domain that is a group. This fact then immediately provides us with a fundamental method of identifying quotient groups:

THEOREM 2.2. *(First Isomorphism Theorem) Let $\varphi : G \longrightarrow H$ be a group homomorphism which is surjective, $N = \ker\varphi$. Then G/N is isomorphic to H by the map $\bar{\varphi}$ which sends the coset $\bar{a} = aN$ to $\varphi(a)$:*

$$\bar{a} = aN \longmapsto \varphi(a).$$

PROOF. Since φ is assumed to be surjective, the fibres are necessarily all non-empty, and these non-empty fibres induces a partition G, consisting of cosets of N. Thus, G/N consists of cosets, i.e., non-empty fibres. Thus the map $\bar{\varphi}$ is the one defined in (2.5), and this map is well-defined; hence it maps bijectively onto $im(\varphi) = H$. Then by construction this map is automatically a group homomorphism, because of φ. □

EXAMPLE 2.45. *Let $\varphi : \mathbb{C}^\times \longrightarrow \mathbb{R}^\times$ be the absolute value map: $a \longmapsto |a|$. This map is a surjective group homomorphism, and its kernel $\ker\varphi$ is the unit circle U; hence the quotient group $\mathbb{C}^\times/U$ is isomorphic to $\mathbb{R}^\times$.*

EXAMPLE 2.46. *Let $\det : GL_n(\mathbb{R}) \longrightarrow \mathbb{R}^\times$ be the determinant function. This map is a surjective group homomorphism, and its kernel is* *the special linear group* $SL_n(\mathbb{R})$*, consisting of those* $n \times n$ *invertible matrices with determinant* 1*. Thus, the quotient group* $GL_n(\mathbb{R})/SL_n(\mathbb{R})$ *is isomorphic to* $\mathbb{R}^\times$*.*

THEOREM 2.3. *(Mapping property of quotient groups) Let* N *be a normal subgroup of a group* G*,* $\pi : G \longrightarrow \bar{G} = G/N$ *the canonical group homomorphism defined by sending* $a \longmapsto \bar{a} = aN$*. If* $\varphi : G \longrightarrow H$ *is a group homomorphism with kernel* K *such that* $N \subseteq K$*, then there exists a* unique *group homomorphisms* $\bar{\varphi} : G/N \longrightarrow H$ *such that* $\bar{\varphi} \circ \pi = \varphi$*:*

$$\begin{array}{ccc} G & \xrightarrow{\pi} & \bar{G} = G/N \\ & \searrow & \downarrow \bar{\varphi} \\ & & H \end{array} .$$

This map is defined by sending $\bar{a} = aN \longmapsto \varphi(a)$*.*

PROOF. We first show that this map is well-defined: if $aN = bN$, then we clearly have $b \in aN$; hence $b = an$ for some $n \in N \subseteq K$; hence $\varphi(b) = \varphi(a)\varphi(n) = \varphi(a) \cdot 1 = \varphi(a)$. Then this map is clearly a group homomorphism because $\bar{\varphi}(aN)\bar{\varphi}(bN) = \varphi(a)\varphi(b) = \varphi(ab) = \bar{\varphi}(abN) = \bar{\varphi}(aNbN)$. Moreover, the defined map making the diagram commutative is automatically unique, because π is surjective. □

Exercises for Chapter 2.11

EXERCISE 2.81. *Prove that the quotient group* $4\mathbb{Z}/20\mathbb{Z}$ *is isomorphic to* $\mathbb{Z}_5$*.*

EXERCISE 2.82. *Prove that the quotient group* $\mathbb{Z}_6/3\mathbb{Z}_6$ *is isomorphic to* $\mathbb{Z}_3$*.*

EXERCISE 2.83. *Prove that if* S *is a cyclic subgroup of order* n *in the dihedral group* D_n *of order* $2n$ *(Definition 3.6), then* D_n/S *is isomorphic to* $\{\pm 1\}$*.*

EXERCISE 2.84. *List all quotient groups of* D_4*.*

EXERCISE 2.85. *Consider* S_4*.*

(1) *Show that the Klein 4 group,* $V = \{1, (12)(34), (13)(24), (14)(23)\}$*, is a normal subgroup of* S_4*.*
(2) *Prove that* S_4/V *is isomorphic to* S_3*.*
(3) *Prove that* A_4/V *is isomorphic to* $\mathbb{Z}_3$*.*

EXERCISE 2.86. *If* N *is a normal subgroup of a group* G *and* $G/N \simeq \mathbb{Z}_5$ *and* $N \simeq \mathbb{Z}_2$*, prove that* G *is abelian.*

EXERCISE 2.87. *Let* $H = \{\pm 1, \pm i\}$ *be the subgroup of* $\mathbb{C}^\times$ *of 4-th root of unity. Describe all cosets of* H *in* G*, and prove that* G/H *is isomorphic to* $\mathbb{C}^\times$*.*

EXERCISE 2.88. *Find all normal subgroups* N *of the quaternion group* H*, and identify the quotients* H/N*.*

EXERCISE 2.89. *Prove that* $(G \times G')/(G \times \{1'\})$ *is isomorphic to* G'*, where* G *and* G' *are groups.*

EXERCISE 2.90. *If two subgroups* $N \subseteq M$ *of a group* G *are normal in* G*, prove that*

$$(G/N)/(M/N) \simeq G/M.$$

EXERCISE 2.91. *Let* $\varphi : G \longrightarrow G'$ *be a surjective homomorphism of groups and let* T *be a subgroup of* G'*. Prove that* T *is normal in* G' *if and only if* $\varphi^{-1}(T)$ *is normal in* G*.*

CHAPTER 3

Toward Abstract Symmetry

History tells us that Galois (1811 – 1832) was the first mathematician who recognized the importance of symmetry on the set of mathematical objects, nowadays called Galois groups associated to a finite field extension of a given field F, consisting of certain automorphisms *fixing the base field* F, and it was he who provided concrete examples and consequently this urged others to the study on the group theory. It is known that Dedekind (1831 – 1916), the last Ph.D. student of Gauss, coined the official definition of a group. Thus, the usefulness of the concept of group was first recognized through an abstract symmetry, but since symmetry is a common phenomenon in virtually every branch of sciences, the study of symmetry is still one of the most appealing application of Theory of Groups. In this chapter, we will learn how to associate certain groups to a given geometric figure with symmetry of simple kinds, and we will apply those results we have learned so far, for instance, the counting formula in Lemma 2.27, in order to have a better understanding of the subject.

1. Group of Symmetries

In Linear Algebra, we have defined that a map from $\mathbb{R}^n$ to itself is a **rigid motion** or **isometry** if it is distance-preserving:

DEFINITION 3.1. *A* ***rigid motion*** *or* ***isometry*** *of* $\mathbb{R}^n$ *is a map* $m\colon \mathbb{R}^n \longrightarrow \mathbb{R}^n$ *which is distance preserving, that is, it is a map satisfying the following condition: If* X, Y *are two points of* $\mathbb{R}^n$*, then the distance from* X *to* Y *is equal to that from* $m(X)$ *to* $m(Y)$*:*

$$|m(X) - m(Y)| = |X - Y|\,. \tag{3.1}$$

Such a rigid motion carries a triangle to a congruent triangle, and therefore it preserves angles and shapes in general.

One easily checks that the set of all rigid motions of $\mathbb{R}^n$ form a group whose law of composition is composition of functions.

DEFINITION 3.2. *The set of all rigid motions of $\mathbb{R}^n$ is called the* ***group M_n of motions.***

The order of M_n is usually not finite. But in this group, we can have many subgroups of finite order if we consider symmetries of a subset F of $\mathbb{R}^n$:

DEFINITION 3.3. *If a rigid motion m carries a non-empty subset F of $\mathbb{R}^n$ to itself, we call it a* ***symmetry*** *of F. That is, a symmetry of a non-empty subset F of $\mathbb{R}^n$ is a rigid motion of $\mathbb{R}^n$ which fixes F.*

Then one also easily checks that the set of all symmetries of F forms a subgroup of M_n, called the **group of symmetries** of the figure F.

From now to the end of this section, we consider symmetries of a subset F in the plane $\mathbb{R}^2$. In order to associate "symmetries" of a geometric figure of $\mathbb{R}^2$ with subgroups of M_2, we first introduce several definitions for symmetries in the plane:

DEFINITION 3.4. *Let F be a subset of the plane. If there is a line l such that two halves of a whole figure F are each other's mirror images, then we say that F has* ***bilateral symmetry, mirror symmetry, mirror-image symmetry, reflection symmetry,*** *or even simply,* ***flip.***

EXAMPLE 3.1. *Here is a typical example of a "geometric" figure F, a regular pentagonal figure, with bilateral symmetry:*

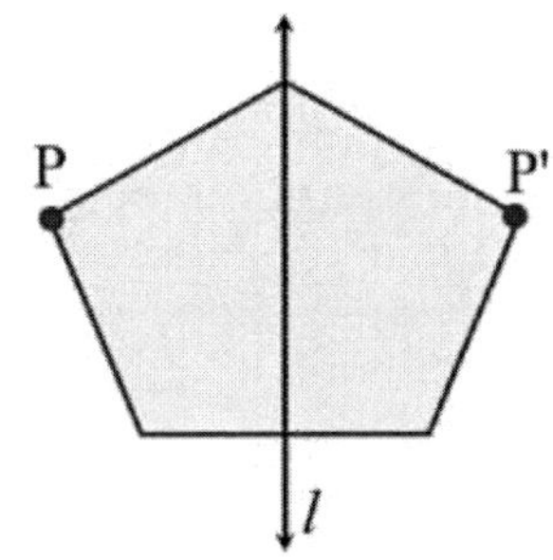

Thus, the group of symmetries associated to this bilateral symmetric figure, with respect to the line l, consists of two elements: the identity transformation 1 and the reflection r about a line called the ***axis of reflection.*** *Thus, it is a cyclic group of order 2.*

DEFINITION 3.5. *Let F be a subset of the plane. If there is an angle θ such that the rotation by θ fixes F, then we say that F has* ***rotational symmetry.***

EXAMPLE 3.2. *The regular pentagonal figure in the last example is fixed under the rotation by $\theta = 72°$, with respect to the center of the figure.*

Thus the group of symmetries associated to this rotational symmetric figure is a cyclic group of order 5.

Then since there is no other symmetry of the figure on the regular pentagonal figure other than those two in the above examples, we can conclude that the group G of symmetries of the regular pentagonal figure *is generated by* these two elements, reflection and rotation:

$$G = \{x, y \in G : x^5 = 1, y^2 = 1, yx = x^{-1}y\},$$

and such a group is called a dihedral group D_5 of order $2 \cdot 5 = 10$. More generally, we have:

DEFINITION 3.6. *The* ***dihedral group*** D_n *of order* $2n$ *is a group generated by two elements* x, y *such that*

$$x^n = 1, \qquad y^2 = 1, \qquad yx = x^{-1}y.$$

Thus, this group consists of elements of the form:

$$D_n = \{x^i y^j : 0 \leq i < n, 0 \leq j < 2\},$$

satisfying a *non-commutative relation* $yx = x^{-1}y$ if $n \geq 3$.

On the other hand, we have:

DEFINITION 3.7. *Let* F *be a subset of the plane. If there is a point* p *such that* $a + p \in F$ *for each* $a \in F$*, this rigid motion is called a* ***translation*** *by* a*, and the figure* F *is said to have* ***translational symmetry****.*

EXAMPLE 3.3. *A typical example of a (non-geometric) figure* F *with translational symmetry is the set* $\mathbb{Z}$ *of integers on the real line, called the* 1-dimensional lattice.

Thus, the group G *of symmetries of this translational symmetric figure is an infinite cyclic group generated by the rigid motion which carries* $F = \mathbb{Z}$ *one unit to the left, and we call such a motion a* ***translation*** t*:*

$$G = \{..., t^{-2}, t^{-1}, 1, t, t^2, ...\}.$$

We note that *if two groups have the same number of generators and the same relations among the generators, they are isomorphic.*

EXAMPLE 3.4. *Let* $F = \{1, 2, 3\}$ *be a subset of* $\mathbb{Z}$*. Then the set of permutations of* F *fixes* F*; we know that they form a group,* S_3*, the symmetric group of order* $3! = 2 \cdot 3$*. Then by Lemma 2.5, we already know that*

$$S_3 = \{x^i y^j : 0 \leq i < 3, 0 \leq j < 2\}.$$

We have already verified that it is generated by two elements of order 2 *and* 3*, respectively satisfying*

$$yx = x^{-1}y.$$

On the other hand, the dihedral group D_3 of order $2 \cdot 3$ is, by definition, generated by two elements satisfying exactly the same relation. Therefore, $D_3 \simeq S_3$, the smallest non-abelian group.

COROLLARY 3.1. *The dihedral group D_3 and the symmetric group S_3 are isomorphic.* □

REMARK 3.1. *For $n > 3$, the dihedral groups D_n and symmetric groups S_n are certainly not isomorphic, because they have different orders, $2n$ and $n!$, respectively.*

To have a little bit of better understanding on the finite subgroups of M_2, we introduce a

NOTATION 3.1. *Let $\mathbf{O}$ denote the group of orthogonal operators on $\mathbb{R}^2$.*

Then we recall that

DEFINITION 3.8. *Left multiplication by an orthogonal matrix, i.e., a square matrix whose inverse is its transpose, is also called an **orthogonal operator**.*

We have proved the following proposition when we study my book, Linear Algebra:

PROPOSITION 3.1. *Let $m\colon \mathbb{R}^n \longrightarrow \mathbb{R}^n$. Then the following statements on the map m are equivalent to each other:*

- **a:** *m is a rigid motion which fixes the origin.*
- **b:** *m preserves dot product; that is, for all $X, Y \in \mathbb{R}^n$,*

$$m(X) \cdot m(Y) = X \cdot Y.$$

- **c:** *m is left multiplication by an orthogonal matrix.* □

Thus, $\mathbf{O}$ consists of those rigid motions of the plane fixing the origin; hence it contains the rotations about the origin and reflections about line through the origin.

THEOREM 3.1. *Let G be a finite subgroup of the group* $\mathbf{O}$ *of* rigid motions of the plane, which fixes the origin. *Then G is one of the following two groups:*

(1) $G = C_n$*: the* cyclic group *of order n, generated by the rotation ρ_θ with $\theta = 2\pi/n$.*
(2) $G = D_n$*: the dihedral group of order $2n$, generated by the rotation ρ_θ with $\theta = 2\pi/n$ and a reflection r about a line through the origin satisfying a relation $r\rho_\theta = \rho_\theta^{-1}r$.*

PROOF. We note that the elements of $\mathbf{O}$ are either the rotations or the reflections, in general. We then have two exclusive cases:

Case 1: All elements of G are rotations. In this case, we prove that G is cyclic of order n. If $G = \{1\}$, then $G = C_1$. Otherwise, G contains a non-trivial rotation ρ_θ. Let θ be the smallest positive angle of rotations among the elements of G, ρ_α be any rotation in G. Then by our choice of θ, $\alpha > \theta$; by division α by θ, we have

$$\alpha = n\theta + \beta,$$

where $n\theta$ is the largest integer multiple of θ such that $n\theta < \alpha$ and $0 \leq \beta < \theta$. If $\beta > 0$, then since ρ_θ and ρ_α are in G, so is $\rho_\beta = \rho_\alpha\rho_{-n\theta}$, which is impossible by our choice of θ. Thus, G is generated by ρ_θ.

Case 2: G contains a reflection. We may adjust coordinates so that the reflection r with respect to the x-axis is in G. Let H be the subgroup of G consisting of all rotations in G. Then H is a cyclic of order n generated by ρ_θ by the same argument done in Case 1. Thus, G contains $2n$ elements, $\rho_\theta^i, \rho_\theta^i r$, $0 \leq i < n$. Moreover, one can easily verify that

$$r\rho_\theta = \rho_{-\theta}r,$$

and hence G contains the dihedral group D_n as a set. Thus, we are left to prove that $G \subseteq D_n$. For this, we consider an arbitrary element g of G. If g is a rotation, then it is an element of H; hence it is an element of D_n. Otherwise, g is a reflection. We then note that $g = \rho_\alpha r$ for some angle

α. Then since $r \in D_n$ and $\rho_\alpha = \rho_\theta^i$ for some i, $g = \rho_\theta^i r \in D_n$ also. Thus, $D_n = G$. □

COROLLARY 3.2. *If $n \geq 3$, then the dihedral group D_n is the group of symmetries of a regular n-gon, namely, a regular n-sided polygon.*

PROOF. A geometric figure, regular n-gon, obviously has a group of symmetries containing rotations and reflections; hence it must be D_n by Theorem 3.1. □

EXAMPLE 3.5. *$D_1 = \{1, r\}$, so it is a cyclic group C_2 of order 2, where the non-trivial element r is a reflection.*

EXAMPLE 3.6. *$D_2 = \{1, \rho_\pi, r, \rho_\pi r\}$, and hence it has three subgroups of order 2, so it is the Klein four group, the smallest non-cyclic group, though it is an abelian group.*

EXAMPLE 3.7. *In example 3.4, we have already seen that $D_3 \simeq S_3$ of order 6 is the smallest non-abelian group.*

In this section, we have applied the concept of symmetry mainly in the case of *geometric figures* in $\mathbb{R}^2$, except in Example 3.4 where $\{1, 2, 3\}$ is different from the usual geometric objects, like regular triangles or regular pentagons, etc. Starting with the next section, we will develop a theory which will suitably allow us to apply the concept of symmetry to other mathematical objects, for instance, certain automorphisms *fixing the base field F*, in a more general situation.

Exercises for Chapter 3.1

EXERCISE 3.1. *Prove that the set of symmetries of a figure F in the plane is a group.*

EXERCISE 3.2. *Let F be a square or a regular hexagon. Find all symmetries of F in each case.*

EXERCISE 3.3. *Prove that a finite group of rotations of the plane bout the origin is cyclic.*

EXERCISE 3.4. *Describe the six symmetries of a cube with one vertex held fixed, and show that they form under composition a group isomorphic to* S_3.

EXERCISE 3.5. *Show that a cube has exactly* 48 *different symmetries.*

EXERCISE 3.6. *Prove that the group of all symmetries of a regular tetrahedron is isomorphic to* S_4.

EXERCISE 3.7. *In the group of symmetries of a regular hexagon, show that the subgroup mapping a given diagonal onto itself isomorphic to the Klein* 4 *group.*

EXERCISE 3.8. *Describe the following subgroups of* S_4 *and determine the order of each of them:*

(1) *All permutations carrying* $\{1,2\}$ *into the set* $\{1,2\}$.
(2) *All permutations carrying* $\{1,2\}$ *into either* $\{1,2\}$ *or* $\{3,4\}$.

EXERCISE 3.9. *Find* 4 *different subgroups of* S_4 *isomorphic to* S_3 *and nine subgroups of* S_4 *isomorphic to* S_2.

EXERCISE 3.10. *Show that there are at least* 30 *different subgroups of* S_6 *isomorphic to* S_3.

2. Group Actions

We start our discussion with two examples.

EXAMPLE 3.8. *(Complex conjugation) Everybody knows that complex conjugation, which sends*

$$\alpha = a + bi \overset{\sigma}{\longmapsto} \bar{\alpha} = a - bi$$

is a bijective map $\sigma : \mathbb{C} \longrightarrow \mathbb{C}$, *and it is compatible with addition and multiplication:*

$$\overline{\alpha+\beta} = \bar{\alpha} + \bar{\beta}, \qquad \overline{\alpha\beta} = \bar{\alpha}\bar{\beta}, \qquad \bar{1} = 1.$$

Thus, it is an automorphism σ *of the field* $\mathbb{C}$ *of complex numbers which fixes the field* $\mathbb{R}$ *of real numbers. We note that it then enables us to have* *a new symmetry of the set G: Since $\sigma^2 = 1$, we see that the set $\{1, \sigma\}$ is a group G of order 2, and moreover, the set G itself is fixed under the operation of σ, as if the set G were a geometric figure in some $\mathbb{R}^n$ which is fixed by a rigid motion. Namely, if we regard the element σ of G as a mirror image of the element 1 of G with respect to a line l, then the element σ of G "is" a bilateral symmetry of G!*

We note that this description is in fact an algebraic version of a symmetry, corresponding to its counterpart in the Euclidean geometry, a bilateral symmetry of the complex plane $= \mathbb{R}^2$ about the real axis: both descriptions give rise to the same cyclic group of order 2.

We can have another example of this kind:

EXAMPLE 3.9. *Let G be a cyclic group of order 3:*

$$G = \{1, x, x^2\},$$

with $x^3 = 1$. Then we have a non-trivial automorphism σ of G:

$$\begin{aligned} 1 &\longmapsto 1, \\ x &\longmapsto x^2, \\ x^2 &\longmapsto x, \end{aligned}$$

and $\sigma^2 = 1$. Thus, $\{1, \sigma\}$ is a cyclic group of order 2. We note that under this automorphism, the group G itself is automatically fixed, i.e., preserved, as if it were a geometric figure in some space $\mathbb{R}^n$ which is fixed by a rigid motion. In fact, if we regard x as the counterclockwise $120°$ rotation of the set G, then x^2 is clock-wise $120°$ rotation of G, and under this setting σ "is" a bilateral symmetry of G!

These two examples suggest us to have the following interpretation: Given a group G, the set of automorphisms of G form a group, $Aut(G)$ with the law of composition being composition of maps. Since each automorphism of G automatically fixes the set G itself, each *automorphism* should be thought as a *symmetry* of G which is compatible with the "structure" of G, where the structure in this time is the "*group law*" of

G, instead of being a geometric figure with a rigid shape in some n-space $\mathbb{R}^n$.

Thus, the words *automorphism* and *symmetry* are more or less synonymous, except that automorphism is used to describe a permutation of a set which fixes some algebraic structure, while symmetry often refers to a permutation which fixes a geometric structure.

This interpretation then allows us to introduce a new concept, the so-called "*group action*" or "*group operation*", which enables us to describe a symmetry of a (geometric or non-geometric) mathematical object S more efficiently and effectively. We start with a definition.

DEFINITION 3.9. *Let S be a set and G be a group. An (left)* ***action*** *of G on S, or (left)* ***operation*** *of G on S, is a map $. : G \times S \longrightarrow S$ satisfying*

(1) *$1.s = s$ for each $s \in S$.*
(2) *(Associative law) $(gg').s = g.(g'.s)$ for each $g, g' \in G$ and $s \in S$.*

A set S with an action of a group G is sometimes called a ***G-set****.*

EXAMPLE 3.10. *Let G be the cyclic group $\{1, r\}$ of order 2 such that $r^2 = 1$ and S the field $\mathbb{C}$ of complex numbers. Then as we have seen in Example 3.8, G acts on the set $\mathbb{C}$ by the rule $G \times \mathbb{C} \longrightarrow \mathbb{C}$, defined by*

$$r.\alpha = \bar{\alpha}, \text{ and } 1.\alpha = \alpha,$$

where $\bar{\alpha}$ is complex conjugation of α. We note that $1.\alpha = r^2.\alpha = r.(r.\alpha) = r.\bar{\alpha} = \alpha$.

REMARK 3.2. *Suppose that a group G act on a set S. If we fix an element $g \in G$ and consider the effect of $g.s$ for each $s \in S$, it induces* ***left multiplication by*** *g on S:*

$$m_g : S \longrightarrow S, \tag{3.2}$$

defined by sending each

$$s \longmapsto g.s,$$

and this is the way the single element g acts on S. *Moreover, this associated left multiplication m_g is obviously a bijection with the inverse*

$$m_{g^{-1}} : S \longrightarrow S,$$

*where $m_{g^{-1}}$ is left multiplication by g^{-1}. Because of this reason, an action of a group G on a set S is often called an **operation** of G on S.*

Here, one should note that the effect of $g.s$ for a fixed $g \in G$ *and varying* $s \in S$ is, by the abuse of language, simply called left multiplication m_g. *Thus, for instance, if $S = G$, then either $m_g(s) = gs$ (a real left multiplication by g) or $m_g(s) = gsg^{-1}$ (conjugation by g) is also called left multiplication by g! This kind of the abuse of language is used, for instance, in Proposition 3.6 and Proposition 4.9.*

This remark suggests us two definitions, the orbit of an element of S and the stabilizer of an element of S.

DEFINITION 3.10. *Let G be a group acting on a set S. Let s be an element of S. The **orbit** of s in S is the subset of S,*

$$\mathcal{O}_s = \{s' \in S : s' = g.s \text{ for some } g \in G\}.$$

EXAMPLE 3.11. *If G = the group of rigid motions and S = the set of triangles in the plane, then $\mathcal{O}_s$ for a given triangle s is the set of all triangles congruent to s.*

LEMMA 3.1. *The orbits for a group action G on a set S are equivalence classes for the relation defined by*

$$s \sim s' \text{ if and only if } s' = g.s \text{ for some } g \in G.$$

PROOF. $s \sim s$ because $s = 1.s$. If $s' \sim s$ then $s' = g.s$; hence $s = g^{-1}.s'$, thus $s \sim s'$. Now, if $s' \sim s$ and $s'' \sim s'$ then $s' = g.s$ and $s'' = g'.s'$ for some $g, g' \in G$; hence $s'' = g'.s' = (g'g).s$; so $s'' \sim s$. □

Thus, the orbits partitions the set S:

$$S = \coprod_{s \in S} \mathcal{O}_s.$$

Namely, S is a union of disjoint orbits $\mathcal{O}_s$. Thus, the group G operates on S by operating independently on each orbit.

EXAMPLE 3.12. *If $G =$ the group of rigid motions and $S =$ the set of triangles in the plane, then S can be partitioned into congruence classes, the orbits for the actions of G on S, and a rigid motion $m \in G$ permutes each congruence class separately.*

In particular,

DEFINITION 3.11. *Let a group G act on a set S. If S consists of only one orbit, then we say that G* ***operates transitively*** *on the set S.*

EXAMPLE 3.13. *The group M of rigid motions acts transitively on the set of points or the set of lines in the plane, but not on the set of triangles, for non-congruent triangles, for instance similar triangles, are abundant in the plane.*

Here is another concept:

DEFINITION 3.12. *Let a group G act on a set S, s an element of S. The* ***stabilizer*** *of s in S is the subset of G,*

$$G_s := \{g \in G : g.s = s\} \ (= Stab(s)).$$

LEMMA 3.2. *Let G act on a set S. Then the stabilizer of s is a subgroup of G.*

PROOF. Obvious. □

Thus, we have

$$g.s = g'.s \text{ if and only if } g'^{-1}g \in G_s.$$

EXAMPLE 3.14. *The stabilizer of the origin of the plane under the action of the group M of rigid motions is the group* $\mathbf{O}$ *of orthogonal operators on $\mathbb{R}^2$, which we have learned when we study my book, Linear Algebra.*

EXAMPLE 3.15. *If the group M of rigid motions acts on the set S of triangles in the plane and if $\triangle$ denotes an equilateral triangle, then the stabilizer $M_\triangle$ of $\triangle$ is its* group of symmetries, *which is isomorphic to D_3, by Corollary 3.2.*

REMARK 3.3. *We note that when we say that a rigid motion m fixes or stabilizes a triangle $\triangle$ it just means that the motion m carries $\triangle$ to itself in permuting the set of triangles, not meaning that m fixes the points of $\triangle$.*

Exercises for Chapter 3.2

EXERCISE 3.11. *Let a group G acts transitively on a set X. For any $x, y \in X$, prove the stabilizer of x is conjugate to the stabilizer of y.*

EXERCISE 3.12. *Determine $Aut(C_4)$, $Aut(C_6)$.*

EXERCISE 3.13. *What are the stabilizers of elements of S_n? Show that they are all conjugate in S_n.*

EXERCISE 3.14. *Prove that the field $\mathbb{Q}$ of rational numbers has no non-trivial automorphism.*

3. The Action on Cosets

Now let H be a subgroup of a group G. Then by Corollary 2.7 we know that the left cosets form a partition of the group G.

DEFINITION 3.13. *We will call the set of left cosets the* **coset space** *and will often denoted by*

$$G/H,$$

although G/H is not *a group* unless *H is a normal subgroup of G.*

Here, the whole point is that G acts on the set G/H in a natural way, regardless of whether it has a group structure or not, by the rule: if aH is a coset in G/H, then the rule is defined by

$$g, aH \longmapsto g.aH := gaH = \{gah : h \in H\},$$

for each $g \in G$. This induced action is well-defined because $aH = bH$ iff $b^{-1}a \in H$ holds. Then we clearly have:

LEMMA 3.3. *This is an action of G on the coset space G/H, operating transitively, and the stabilizer of H is the subgroup H of G.*

PROOF. It is clearly an action; and it operates transitively because G/H is an orbit of $H = 1H$. The last assertion is true because H closed under multiplication. □

EXAMPLE 3.16. *Let $G = D_3$, the group of symmetries of an equilateral triangle. Then by Corollary 3.2, it is described by two generators x, y with three relations between x and y: $x^3 = 1, y^2 = 1, yx = x^2y$. Let $H = \{1, y\}$ be a subgroup of G. Then we have three cosets of H:*

$$G/H = \{H, xH, x^2H\},$$

and D_3 operates on G/H. Thus, every element $g \in D_3$ determines a permutation m_g of G/H. For instance,

$$x : \begin{array}{ccc} H & \mapsto & xH \\ xH & \mapsto & x^2H \\ x^2H & \mapsto & H \end{array},$$

and

$$y : \begin{array}{ccc} H & \mapsto & H \\ xH & \mapsto & x^2H \\ x^2H & \mapsto & xH \end{array}.$$

One checks in this way that the six elements of D_3 induce all six permutations of G/H; hence the map: $G = D_3 \longrightarrow S_3 = Perm(G/H)$ defined by sending for each $g \in D_3$

$$g \longmapsto m_g$$

is an isomorphism of groups, which we have already seen before in Example 3.4.

More generally, the following proposition relates an arbitrary group action to the operations on cosets in terms of the stabilizer of an element s and the orbit of s:

PROPOSITION 3.2. *Let a group G act on a set S, s an element of S, G_s the stabilizer of s, and $\mathcal{O}_s$ the orbit of s. Then there exists a "natural" bijective map $\varphi : G/G_s \longrightarrow \mathcal{O}_s$ defined by*

$$aG_s \longmapsto a.s,$$

where the map is "natural" in the sense that it satisfies

$$\varphi(gaG_s) = g.\varphi(aG_s),$$

for each coset $aG_s \in G/G_s$ and each $g \in G$.

PROOF. Once such a map φ exists, then

$$gaG_s \longmapsto (ga).s = g.(a.s) = g.\varphi(aG_s),$$

hence φ is *compatible with the operations* of G on the set G/G_s.

We first need to show that φ defined such is well-defined: For this, suppose that $aG_s = bG_s$. Then $a^{-1}b \in G_s$, or equivalently, $b = ah$ for some $h \in G_s$; hence $b.s = (ah).s = a.s$ because G_s fixes s. We have shown that the same coset with different representatives a and b, respectively, are map to the same image, so φ is a well-defined map.

For any $g.s \in \mathcal{O}_s$, we have $gG_s \in G/G_s$ such that $\varphi(gG_s) = g.s$. Thus, φ is surjective.

Now finally suppose that $\varphi(aG_s) = \varphi(bG_s)$. Then $a.s = b.s$; hence $(a^{-1}b).s = s$; hence $a^{-1}b \in G_s$; hence $b = ah$ for some $h \in G_s$; hence $b \in aG_s$; hence $aG_s = bG_s$. This proves that φ is injective. □

REMARK 3.4. *Proposition 3.2 says that every group action G on a set S can be described in terms of the operations on the cosets. More precisely, this is what it means: Since S is partitioned as a union of disjoint orbits $\mathcal{O}_s$, and since each orbit $\mathcal{O}_s$ is naturally bijective to the set of cosets G/G_s of the stabilizer G_s of s of which the orbit is $\mathcal{O}_s$, the effect of the original group action on S is exactly the same as group actions of G on various sets of cosets G/G_s, where each G_s is determined by each s of which orbit*

$\mathcal{O}_s$ is the disjoint component of the partition of S:

$$S = \coprod_{s \in S} \mathcal{O}_s.$$

In other words, if N is the number of disjoint $\mathcal{O}_s$'s, then there should be N number of G_s's and hence N number of sets G/G_s on each of which the group G operates, so that the effect of G operating on N different sets G/G_s of cosets with respect to N different G_s's, simultaneously or separately, is exactly the same as G operating on S just once.

In particular, if G happens to act on S transitively, then there is only one orbit and hence only one set of cosets G/G_s.

EXAMPLE 3.17. *We consider Example 3.16 again. Let $S = \{v_1, v_2, v_3\}$ be the set of vertices of an equilateral triangle. Since y is a reflection, it fixes one vertex, say v_1. Thus, the stabilizer of v_1 is $H = \{1, y\}$. Then since H is a subgroup of a group G, the group G acts on G/H transitively by Lemma 3.3, where G/H consists of H, xH, and x^2H. Thus by Proposition 3.2, we have a natural bijection:*

$$\begin{aligned} H &\longmapsto v_1, \\ xH &\longmapsto v_2, \\ x^2H &\longmapsto v_3. \end{aligned}$$

With the following proposition, we finish this section.

PROPOSITION 3.3. *Let a group G act on a set S, s an element of S, G_s the stabilizer of s, and let s' be an element of the orbit $\mathcal{O}_s$ of s so that*

$$s' = a.s$$

for some $a \in G$. Then

(1) *The the following two subsets of G are equal:*

$$\{g \in G : g.s = s'\} = aG_s,$$

where

$$aG_s := \{g \in G : g = ah \text{ for some } h \in G_s\}.$$

(2) *The stabilizer* $G_{s'}$ *of* s' *is a* conjugate subgroup *of the stabilizer* G_s *of* s*:*

$$G_{s'} = aG_sa^{-1}, \tag{3.3}$$

where

$$aG_sa^{-1} := \{g \in G : g = aha^{-1} \text{ for some } h \in G_s\}.$$

PROOF. Part (1). For any g such that $g.s = s'$, $g.s = a.s$; hence their preimages under the bijective natural map φ defined in Proposition 3.2 must be the same: $gG_s = aG_s$; hence $g = g1 = ah$ for some $h \in G_s$. Now, if $g \in aG_s$ then $gs = ahs$ for some $h \in G_s$; hence $g.s = a.s = s'$. This establish the equality.

Part (2). If $g \in G_{s'}$ then $g^{-1} \in G_{s'}$; hence $g^{-1}.s' = s' = a.s$; hence $s' = ga.s$; hence $ga \in aG_s$ by Part (1). Thus $g \in aG_sa^{-1}$. Now, if $g \in aG_sa^{-1}$ then $ga \in aG_s$; hence $ga.s = s'$ again by Part (1). Thus, $g.s' = s'$; hence $g \in G_{s'}$. This establish the equality. □

COROLLARY 3.3. *Under the same setting as in Proposition 3.3, the order of* G_s *is the same as that of the set* $\{g \in G : g.s = s'\}$*.*

PROOF. This is because $|G_s| = |aG_s|$. □

EXAMPLE 3.18. *This is a continuation of Example 3.17. Since* $H = \{1, y\}$ *is the stabilizer of* v_1*,* $xH = \{x, xy\}$*.*

EXAMPLE 3.19. *This is a continuation of Example 3.17. Since* H *is the stabilizer of* v_1*, by part (2) of Proposition 3.3,* xHx^{-1} *is the stabilizer of* $v_2 = xv_1$*.*

REMARK 3.5. *In the next section, we shall see that we have in fact proved an important equality (3.3) in Proposition 3.3 for finite groups, which connects a group action on itself by left multiplication to a conjugation relation among subgroups of* G*. For example, we need (3.3) in Proposition 3.3 in the proof of Second Sylow Theorem, Theorem 4.7.*

Exercises for Chapter 3.3

EXERCISE 3.15. *For dihedral groups D_3 and D_6,*

1. *Show that D_3 is a subgroup of D_6.*
2. *Determine all conjugate subgroups of D_3 in D_6, and describe the cosets of D_3.*

4. The Counting Formulas

We first review Lemma 2.27. For this let H be a subgroup of a group G. Then by Part (1) of Lemma 2.27, all the cosets of H in the coset space G/H have the same cardinality:

$$|H| = |aH|.$$

Then since G is a union of disjoint cosets and the number of cosets is the index,

$$[G : H] = |G/H|,$$

we get the "fundamental" counting formula for the order $|G|$ of G:

$$|G| = |H| \cdot |G/H|, \tag{3.4}$$

which is Lemma 2.27.

If we combine this result with Proposition 3.2 with a general G-set S, we obtain:

PROPOSITION 3.4. *Let a group G act on a set S, and let $s \in S$ be an element of S. Then*

$$|G| = |G_s| \cdot |\mathcal{O}_s|, \tag{3.5}$$

namely, the order of G is equal to the order of the stabilizer G_s times the order of orbit $\mathcal{O}_s$. Or equivalently, the order of the orbit is equal to the index of the stabilizer. □

COROLLARY 3.4. *The order of an orbit divides the order of the group G.* □

On the other hand, we know that, under the action of a group G on a set S, S is partitioned into orbits, so if we label the different orbits by $\mathcal{O}_i$, we have a more "elementary" **counting formula**:

$$|S| = |\mathcal{O}_1| + |\mathcal{O}_2| + \cdots .$$

In particular, if S is a finite set, then

$$|S| = |\mathcal{O}_1| + |\mathcal{O}_2| + \cdots + |\mathcal{O}_k| , \tag{3.6}$$

for some $k \in \mathbb{N}$. Here, we note that the cardinalities $|\mathcal{O}_i|$ of $\mathcal{O}_i$'s are not the same in general.

In counting business, these simple formulas have many applications and here is just one of them:

EXAMPLE 3.20. *Let G be the icosahedral group I, namely, the group of orientation-preserving symmetries of a regular dodecahedron D:*

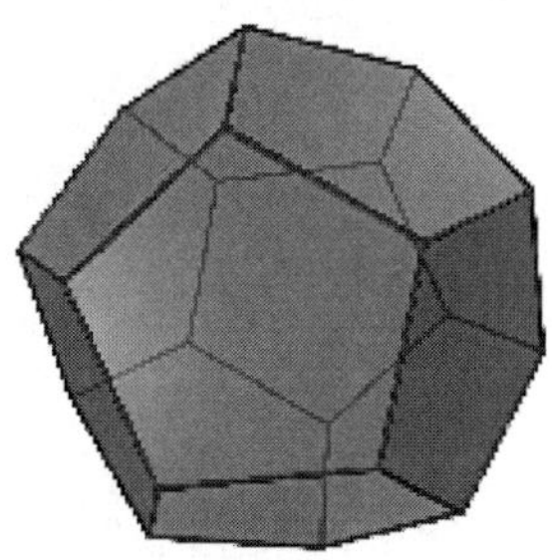

Dodecahedron

Thanks to the above rather transparent drawing of D, we can easily count all vertices, edges, and faces of D. D has 12 faces, 30 edges, 20 vertices, and each vertex of D has three adjacent faces.

In what follows, we may assume that this geometric figure in $\mathbb{R}^3$ is centered at the origin of $\mathbb{R}^3$. Then, the symmetries of D, i.e., the elements of G, are all rotations fixing the origin. We are going to count all the symmetries of D.

We first consider the group action of G on the set S of the faces of D. The stabilizer G_s of a face s is the group of rotations by multiples of $2\pi/5$ about a perpendicular line through the center of the face s. Thus,

G_s consists of 5 elements. Then G operates transitively on the coset space G/G_s, and hence we have:

$$|G| = |G_s|\,|G/G_s| = 5 \cdot 12 = 60,$$

where $|G/G_s| = 12$ is the number of faces which we already know.

Similarly, G operates transitively on the set of 20 vertices v of D, and if v is a vertex of D, then G_v consists of three rotations including the identity map, and hence $|G_v| = 3$. Then since there are 20 vertices, we also obtain: $|G| = 3 \cdot 20 = 60$.

Similarly, if we denote an edge of D by e, then the stabilizer G_e of e has two elements including the identity map, and hence $|G_e| = 2$; consequently, $|G| = 2 \cdot 30 = 60$.

REMARK 3.6. *One important aspect of a group action on a set is that if G acts on a set S, then a subgroup H of G also acts on S. In particular, for an element $s \in S$, the H-orbit of s is obviously contained in G-orbit of s; consequently we can decompose a single G-orbit of s into a union of H-orbits, and this allows us to have another counting formula for the orbit associated to the stabilizer of s, which is a variant of the elementary counting formula in (3.6).*

EXAMPLE 3.21. *(This example is a continuation of Example 3.20.) Let S be the set of 12 faces of the dodecahedron D, H the stabilizer G_s of a particular face s of D. Since H also fixes the face opposite to s, there are two H-orbits of order 1. The remaining 10 faces make up two orbits of order 5 under the action of $H = G_s$. Thus, by the counting formula in (3.6) applied to this case, we see that*

$$12 = 1 + 1 + 5 + 5.$$

Similarly, let S be the set of faces, K be the stabilizer G_v of a vertex. We already know that $|G_v| = 3$. Let K acts on S. Then since K does not fix any face at all, every K-orbit has order 3; and consequently we see that

$$12 = 3 + 3 + 3 + 3.$$

Here is a simple application of the above remark to the case that G-set is the coset space of a subgroup of a group.

PROPOSITION 3.5. *Let H and K be two subgroups of a group G. Then the index of the subgroup $H \cap K$ in H is at most equal to the index of K in G:*

$$[H : H \cap K] \leq [G : K] .$$

PROOF. Let S be the coset space G/K, s the coset $1K = K$. Then by Lemma 3.3 G acts on S transitively and the stabilizer of s is the subgroup K. We now *restrict* the action of G to the subgroup H, and by doing so we decompose S into a union of disjoint H-orbits. We then easily note that the stabilizer of s for this restricted operation is $H \cap K$. Although we do not know much about the H-orbit $\mathcal{O}_s$ of s except that it is a subset of S, we can still apply Proposition 3.2 and we see that

$$|\mathcal{O}_s| = [H : H \cap K] .$$

Therefore, we conclude that $[H : H \cap K] = |\mathcal{O}_s| \leq |S| = [G : K]$. □

More generally, the following theorem can be proved using the counting formula in Proposition 3.4:

THEOREM 3.2. *Every subgroup G of the special orthogonal group SO_3,*

$$SO_3 = \{A \in GL_3(\mathbb{R}) : A^t A = I, \ \det(A) = 1\},$$

is one of the following:

(1) *The cyclic group C_k of rotations by multiples of $2\pi/k$ about a line;*
(2) *The dihedral group D_k of symmetries of a regular k-gon;*
(3) *The tetrahedral group T of twelve rotations carrying a regular tetrahedron to itself;*
(4) *The octahedral group O of order 24 of rotations of a cube, or of a regular octahedron;*

(5) *The icosahedral group I of 60 rotations of a regular dodecahedron or a regular icosahedron.*

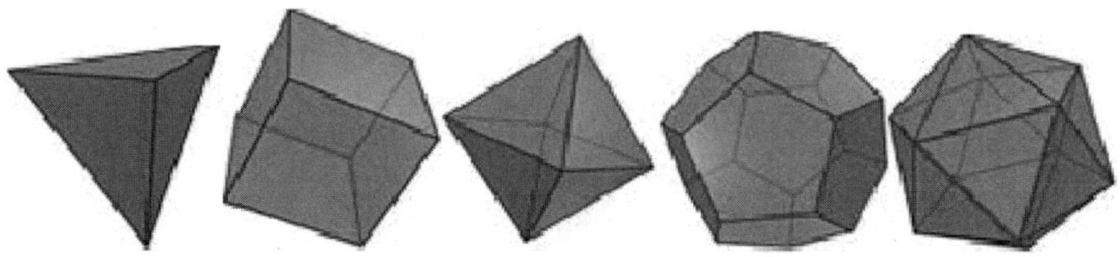

Tetrahedron Hexahedron Octahedron Dodecahedron Icosahedron

REMARK 3.7. *We recall that the geometric figures in $\mathbb{R}^3$, tetrahedron, cube, octahedron, dodecahedron, and icosahedron, are called **Platonic solids**. It is well-known that Euclid proved that they are all regular polyhedra in $\mathbb{R}^3$ in his last book of the Elements. We note that the cube and the octahedron are "dual" to each other, and the dodecahedron and the icosahedron are dual to each other. That is, we obtain the regular icosahedron from the regular dodecahedron by "dually" replacing the words "vertices" with the words "faces". Thus, the tetrahedron is self-dual. We note that tetra- = 4, hexa = 6, octa- = 8, dodeca- = 12, and icosa- = 20, and hence they are called such and such. (Figures taken from: http://en.wikipedia.org/wiki/Platonic_ solid)*

Exercises for Chapter 3.4

EXERCISE 3.16. *Using the counting formula in (3.6), determine the order of the group of rotational symmetries of a cube.*

EXERCISE 3.17. *Using the counting formula in (3.6), determine the order of the group of rotational symmetries of a tetrahedron.*

EXERCISE 3.18. *Compute the order of the group of symmetries of a cube, when orientation-reversing symmetries such as reflections in the planes, as well as rotations, are allowed.*

EXERCISE 3.19. *Compute the order of the group of symmetries of a tetrahedron, when orientation-reversing symmetries such as reflections in the planes, as well as rotations, are allowed.*

EXERCISE 3.20. *Compute the order of the group of symmetries of a dodecahedron, when orientation-reversing symmetries such as reflections in the planes, as well as rotations, are allowed.*

EXERCISE 3.21. *Explain why the groups of symmetries of the cube and octahedron, and of the dodecahedron and icosahedron, are the same, respectively.*

EXERCISE 3.22. *Prove that the icosahedral group has a group of order 10.*

EXERCISE 3.23. *Prove Theorem 3.2.*

5. Permutation Representations

Before we move to the next Chapter of which the starting point is a group action G on G itself, we close this Chapter with a little bit more abstract concept for a better understanding of a given group G.

We recall that the symmetric group S_n, the group of permutations on the set $\{1, 2, ..., n\}$, by its definition, acts on the set $S = \{1, 2, ..., n\}$. This has a subtle generalization:

DEFINITION 3.14. *A group homomorphism $\varphi : G \longrightarrow S_n$ for some n is called a* ***permutation representation*** *of a group G.*

EXAMPLE 3.22. *In the case of S_n, we have a permutation representation with $\varphi =$ the identity map.*

Suppose that a group G acts on any set S. Then for any fixed $g \in G$, we have seen in Remark 3.2 that left multiplication m_g by g:

$$s \longmapsto g.s$$

is a bijection p of S. Then since we have $m_{gh} = m_g m_h$, we see that the map from G to $Perm(S)$ defined by $g \longmapsto m_g$ is a group homomorphism. On the other hand, we note that any group homomorphism $\varphi : G \longrightarrow Perm(S) :=$ the set of all bijections, i.e., permutations, on S gives rise to a group action of G on the set S by letting m_g be the permutation $\varphi(g)$, namely, by defining,

$$g.s = \varphi(g)(s).$$

LEMMA 3.4. *This is a group action of G on S.*

PROOF. We note that $\varphi(1) = 1$ is the identity permutation of S and $\varphi(g)$ is a permutation of S. Thus, for any $s \in S$, we define an action of G on S by letting m_g be the permutation $\varphi(g)$, namely, by defining,

$$g.s := \varphi(g)(s).$$

Then $1.s = \varphi(1)(s) = 1(s) = s$ and

$$(gg').s = \varphi(gg')(s) = \varphi(g)\varphi(g')(s) = \varphi(g)(\varphi(g')(s)) = g.(g'.s).$$

Thus, it defines a group action of G on S through the group homomorphism $\varphi : G \longrightarrow S$. $\square$

More generally, this association is in fact a one-to-one correspondence:

PROPOSITION 3.6. *Let G be a group, S any set. There exists a one-to-one correspondence between the set of group actions of G on S and the set of group homomorphisms*

$$\varphi : G \longrightarrow Perm(S),$$

where $Perm(S)$ denotes the group of permutations of S. The correspondence is defined as follows: given a group action of G on S, φ is defined by the rule,

$$\varphi(g) := m_g,$$

where m_g is left multiplication by g, defined in Remark 3.2.

PROOF. First, we show that the map φ defined in the proposition is a group homomorphism: We have already seen in Remark 3.2 that m_g is a permutation of S; hence $\varphi(g) = m_g$ belongs to $Perm(S)$. Then for any $s \in S$, we have:

$$m_{gh}(s) = (gh).s = g.(h.s) = m_g m_h(s),$$

showing that $\varphi(gh) = m_{gh} = m_g m_h = \varphi(g)\varphi(h)$ for any $g, h \in G$, and this proves that the map φ defined in the proposition is indeed a group homomorphism.

Then, we have, through the definitions $m_g := \varphi(g)$ and $\varphi(g) := m_g$, respectively, that

$$\{\varphi : G \longrightarrow Perm(S)\} \longrightarrow \{\{m_g\}_{g\in G}\} \longrightarrow \{\varphi : G \longrightarrow Perm(S)\},$$

and

$$\{\{m_g\}_{g\in G}\} \longrightarrow \{\varphi : G \longrightarrow Perm(S)\} \longrightarrow \{\{m_g\}_{g\in G}\},$$

are identity maps, respectively, where the set of all group actions G on S is denoted by $\{\{m_g\}_{g\in G}\}$. Thus, they are in one-to-one correspondence. □

REMARK 3.8. *We note that the above proof is in fact very similar to the traditional proof of the so-called* ***Cayley's Theorem****: Every group is isomorphic to a transformation group, namely, a group of permutations. Our version of his theorem is Theorem 4.1.*

EXAMPLE 3.23. *In Example 3.4, we have already seen that the group* D_3 *is isomorphic to* S_3*, which is an typical example of a group representation* φ*.*

In contrast to this example, a general group representation φ needs not be injective or surjective.

DEFINITION 3.15. *If a group representation* $\varphi : G \longrightarrow Perm(S)$ *is injective, we say that the corresponding group action of* G *on* S *is* ***faithful.***

LEMMA 3.5. *The following statements are equivalent to each other:*

(1) φ *is injective.*
(2) m_g *is not the identity map unless* $g = 1$*.*
(3) *If* $g.s = s$ *for* every $s \in S$ *then* $g = 1$*.* □

Here is an example of a group representation which is faithful:

EXAMPLE 3.24. *The action of the group M of rigid motions in the plane, $\mathbb{R}^2$, on the set S of equilateral triangles in the plane is faithful, because the identity map is the only rigid motion which fixes* all *triangles.*

REMARK 3.9. *If G does not act faithfully on a set S, one can easily modify the group to obtain a faithful action. For this, we do the followings: Define*

$$N = \{g \in G : g.s = s \text{ for each } s \in S\}.$$

Then by Exercise 3.26, N is a normal subgroup of G. One then easily verifies that the quotient group G/N acts faithfully on S by defining its action on S by the rule:

$$(gN).s = g.s.$$

We also note that the original action of G on S is faithful if and only if $N = \{e\}$ (Lemma 3.5).

REMARK 3.10. *The main reason why we introduce the concept of a permutation representation of a group action G on a set S is that it allows us to have a more detailed description of certain important groups in terms of permutations, where the arbitrarily choosable set S plays a role of the degree of freedom, so to speak.*

Similarly, to have a better understanding of a given group G, we have other concepts, which are similar to that of permutation representation, in the sense that we know pretty well how to handle or compute them *in an explicit manner, like permutations:*

DEFINITION 3.16. *Let G be a group.*

(1) *An n-**dimensional matrix representation** of G is a group homomorphism $R : G \longrightarrow GL_n(F)$, where F is a field.*
(2) *A **representation** of G on a finite-dimensional vector space V is a group homomorphism $\rho : G \longrightarrow GL(V)$.*

REMARK 3.11. *Since a choice of a basis of a finite-dimensional vector space V defines an isomorphism:*

$$GL(V) \simeq GL_n(F),$$

by sending $T \longmapsto$ the matrix of T with respect to a given basis, a representation of G on a finite dimensional vector space V is a mere generalization of an n-dimensional matrix representation in the sense that matrix representations of a group G are representations of G on the vector space F^n of column vectors.

As we have mentioned in the above remark, we know fairly well about how to handle or compute invertible matrices in an explicit manner. This means that once we have a representation of a group G, we may be able to have a better understanding of G itself from, Algebraic Analysis *or* Symbolic Computation, *i.e., exact computation-oriented experiments.*

Here is an application of permutation representations of a group G, which enable us to have a better understanding of G.

PROPOSITION 3.7. *The map $\varphi : S_3 \longrightarrow Aut(S_3) =$ the group of automorphisms of S_3, defined by conjugation $g \longmapsto c_g$, where*

$$c_g(z) = gzg^{-1},$$

is an isomorphism of groups.

PROOF. By Lemma 2.17, we know that the conjugation c_g is an automorphism of S_3. Then φ is automatically a group homomorphism: $c_{gh}(z) = (gh)z(gh)^{-1} = g(hzh^{-1})g^{-1} = c_g c_h(z)$ for each $z \in S_3$. Then we have, by definition, that conjugation c_g by g is the identity map if and only if g is in the center of S_3. But, already know that the center of S_3 is $\{1\}$ (See Example 2.26). Thus, φ is injective.

The proof that φ is surjective requires some works. In what follows, we are going to prove that we have an *injective* permutation representation

$$\phi : Aut(S_3) \longrightarrow Perm\{y, xy, yx\},$$

where x, y are generators of S_3. Once this is shown, it follows that the order of $Aut(S_3)$ is at most 6, because the order of $Perm\{y, xy, yx\}$ is 6, and then since we already have shown that the map $\varphi : S_3$ of order 6 to $Aut(S_3)$ is injective, φ must be surjective; hence this finishes the whole proof.

For this, first of all, we note that the group S_3 has three elements, y, xy, yx, of order 2 (Corollary 2.1) and S_3 is decomposed into

$$S_3 = \{e\} \cup \{y, xy, yx\} \cup \{x, x^2\},$$

according to the order of its elements. Secondly, we have an obvious group action of $Aut(S_3)$ on S_3 defined by

$$\alpha.s = \alpha(s),$$

for each $s \in S_3$. Then since elements of S_3 of different orders belong to different orbits for this action (to verify this, apply conjugations in $Aut(S_3)$), *we restrict the action of* $Aut(S_3)$ *to the orbit* $\mathcal{O}$ *consisting of those elements of order* 2. Again applying conjugations in $Aut(S_3)$, one can easily verify that this orbit $\mathcal{O} = \{y, xy, yx\}$. Cf. Corollary 2.1. Thus, we have a unique permutation representation $\phi : Aut(S_3) \longrightarrow Perm\{y, xy, yx\}$, associated to this action by Proposition 3.6. Now, if an automorphism α fixes xy and y, then $xyy = x$ is also fixed; hence yx is also fixed by α. In particular, we have seen that if α fixes each element of $\mathcal{O}$, then x and y are fixed. Then since x and y generate S_3, the identity is the *only* such automorphism α. Thus, by Lemma 3.5, $Aut(S_3)$ acts *faithfully* on the orbit $\mathcal{O} = \{y, xy, yx\}$. That is, the associated permutation representation in Proposition 3.6

$$\phi : Aut(S_3) \longrightarrow Perm\{y, xy, yx\}$$

is injective. □

Exercises for Chapter 3.5

EXERCISE 3.24. *Let G be the dihedral group D_4 of symmetries of a square.*

(1) *Prove or disprove that its action on vertices is a faithful action or not.*
(2) *Prove or disprove that its action on diagonals is a faithful action or not.*

EXERCISE 3.25. *Prove Lemma 3.5.*

EXERCISE 3.26. *Let a group G act on a set S. Prove that*

$$N = \{g \in G : g.s = s \text{ for each } s \in S\}$$

is a normal subgroup of G. (Hint: it is the kernel of a certain group homomorphism from G.)

CHAPTER 4

More Group Theory

1. The Actions of a Group on Itself

In the last Chapter, we have studied a group action G on a set S, where S is arbitrary. In this section, we choose $S = G$. Among several possible actions of G on itself, we consider the most important ones: *left multiplication by* g and *conjugation by* g, $c_g(x) = gxg^{-1}$.

1.1. Left Multiplication by g **.** Left multiplication by g, $G \times G \longrightarrow G$,sending

$$g, x \longmapsto g.x := gx, \tag{4.1}$$

is clearly a transitive action of G on itself, because $g' = (g'g^{-1})g$; hence it has a single orbit. Moreover the stabilizer of any element is obviously the identity subgroup $\{1\}$. Thus, the action is faithful by Lemma 3.5, and hence the associated permutation representation

$$G \longrightarrow Perm(G), \tag{4.2}$$

defined in Proposition 3.6 is automatically injective. We have a famous theorem:

THEOREM 4.1. *(Cayley's Theorem) Every finite group G is isomorphic to a subgroup of a permutation group. If G has order n, then it is isomorphic to a subgroup of the symmetric group S_n.*

PROOF. Since the action by left multiplication is faithful, G is isomorphic to its image in $Perm(G)$. If G has order n, then $Perm(G)$ is isomorphic to S_n. □

REMARK 4.1. *As we have already explained in Remark 3.8, Clayey's Theorem is very interesting. But, it is not very useful especially when n is relatively large because the order of S_n is $n!$*

1.2. Conjugation by g . Conjugation by g, $G \times G \longrightarrow G$, sending

$$g, x \longmapsto gxg^{-1}, \tag{4.3}$$

is clearly an action of G on itself.

Because of historical reason, the stabilizer of an element $x \in G$ for the action of conjugation has another name, the **centralizer** of x (the **normalizer** of x) and is denoted by $Z(x)$:

$$Z(x) := \{g \in G : gxg^{-1} = x\} = \{g \in G : gx = xg\} \tag{4.4}$$

Thus, the centralizer is the set of group elements which commute with x. We note that the subgroup $Z(x)$ of G always contains x, because it commute with itself.

Also because of historical reason, the orbit of an element $x \in G$ for the action of conjugation has another name, the **conjugacy class** of x and is denoted by C_x:

$$C_x := \{x' \in G : x' = gxg^{-1} \text{ for some } g \in G\}. \tag{4.5}$$

In particular, we obviously have:

COROLLARY 4.1. *Every element belonging to the same conjugacy class has the same order.* □

REMARK 4.2. *The converse of this corollary does not hold in general. For this, see Subsection 2.1 below.*

Then by the Counting Formula in Proposition 3.4, we have

$$|G| = |Z(x)| \cdot |C_x| \,. \tag{4.6}$$

Then since the conjugacy classes are orbits for a group action, they partition G, and this gives what is usually called the **Class Equation** for a finite group G:

$$|G| = \sum_{\substack{\textit{conjugacy} \\ \textit{classes } C_x}} |C_x| = \sum_{\substack{\textit{conjugacy} \\ \textit{classes } C_x}} \frac{|G|}{|Z(x)|}. \tag{4.7}$$

which is a special version of the equation in (3.6). If we number the conjugacy classes, as C_i, $i = 1, 2, ..., k$, then this formula reads

$$|G| = |C_1| + |C_2| + \cdots + |C_k|.$$

In particular, the conjugacy class of the identity element 1 of G must appear in the right side of this equation, and this conjugacy class consists of 1 alone, because the identity element 1 is left fixed by all $g \in G$: $g1g^{-1} = 1$. Thus, we denote this conjugacy class by

$$C_1 = \{1\}.$$

Since each conjugacy class C_i is an orbit, its order $|C_i|$ divides the order $|G|$ of the group G (Proposition 3.4), so that each term in the right side of the equation in (4.7) divides $|G|$. Thus, we have proved:

COROLLARY 4.2. *Each of the numbers on the right side of the Class Equation in (4.7) divides the order of* G, *and* at least *one of them is equal to* 1. □

Later in this section, we will see more than one of them are equal to 1 under what condition.

EXAMPLE 4.1. *The conjugacy classes in the symmetric group* $S_3 \simeq D_3$, *the dihedral group in Proposition 3.1 and Example 3.4, consist of the following three subsets:*

$$\{1\}, \{x, x^2\}, \{y, xy, x^2y\}.$$

This is so, because we have:

$$\begin{aligned} yxy^{-1} &= x^{-1}yy^{-1} = x^2, \quad yx^2y^{-1} = yx^{-1}y = yyx = x; \\ xyx^{-1} &= x^{-2}yx^{-1} = x^{-1}yxx^{-1} = x^2y, \quad x^{-1}yx = x^2yx = x^2x^{-1}y = xy. \end{aligned}$$

Thus, we have explicitly verified that the two rotations x, x^2 of the same order are conjugate to each other, and the three reflections y, xy, x^2y of the same order are conjugate to each other. Thus, the Class Equation for D_3 is

$$6 = 1 + 2 + 3.$$

REMARK 4.3. *In the above example, we happen to see that elements of the same order belong to the same conjugacy class. But, this phenomenon is* not *the case in general, as we shall see it in the next Section in the case of the icosahedral group I of order* 60 *of orientation-preserving symmetries of a regular dodecahedron D. [See (4.10) below].*

1.3. Center of a Group. We now recall that the **center** Z of a group G is:

$$Z = \{g \in G : gx = xg \text{ for } each\ x \in G\}.$$

Thus, Z is a subgroup of the centralizer $Z(x)$ of *any* $x \in G$. In fact, we have

$$Z = \bigcap_{x \in G} Z(x). \tag{4.8}$$

Moreover, Lemma 2.21 says that Z is an abelian subgroup of G. Furthermore, we have the following technical criterion:

PROPOSITION 4.1. *Let x be an element of a group G. Then the conjugacy class C_x of x consists of x alone, namely, its centralizer $Z(x)$ is the whole group G, if and only if $x = gxg^{-1}$ for each $g \in G$ if and only if $x \in Z$.*

PROOF. Just apply the involved definitions. □

Thus, we have:

COROLLARY 4.3. *Those elements x of the center Z of a group G are all represented by the* 1*'s on the right side of the Class Equation, and conversely.* □

There is a particular case in which the Class Equation in (4.7) shines, namely when G is a p-group.

DEFINITION 4.1. *A group G is called a p-**group** if its order is a positive power p^r of a prime p.*

In the rest of this section, p is a prime number and we have a few applications of the Class Equation in (4.7):

THEOREM 4.2. *(Burnside) Let p be a prime number. Then the center of a p-group G has order > 1.*

PROOF. We need to prove that the center Z of G, which is an abelian subgroup of G, has a non-identity element x. Since G is a p-group, we first note that both terms in the left side and in the right side of the Class Equation in (4.7) are all powers of p. Thus, if $C_1 = \{1\}$ in the Class Equation in (4.7) is the only conjugacy class consisting of a single element, then we fall into the contradiction that 1 is divisible by p. □

COROLLARY 4.4. *The order of the center Z of a p-group G is at least p.* □

PROPOSITION 4.2. *(Fixed Point Theorem for a p-group action on a finite set S) Let p be a prime number and let a p-group G act on a finite set S. If the order of S is not divisible by p, then there exists a fixed point $s \in S$ for the action of G on S. Namely, there exists an element $s \in S$ of which stabilizer is the whole group G.*

PROOF. In the elementary counting formula,

$$|S| = |\mathcal{O}_1| + |\mathcal{O}_2| + \cdots + |\mathcal{O}_k|,$$

in (3.6), each $|\mathcal{O}_i|$ is either 1 or divisible by p because G is a p-group. Then since $|S|$ is not divisible by p, the number N of those $\mathcal{O}_i$ of order 1 must not be a multiple of p and this $N \geq 1$. □

PROPOSITION 4.3. *Let p be a prime number. Then every group G of order p^2 is abelian.*

PROOF. If we can prove that the centralizer $Z(x)$ of each $x \in G$ is the whole group G, then we are done. For this, let $x \in G$ be *any* element. If

$x \in Z$, the center of G, then $Z(x) = G$ by Proposition 4.1. Otherwise, $x \notin Z$; hence the subgroup $Z(x)$, which always contains x, properly contains Z by (4.8):

$$\{1\} \subsetneqq Z \subsetneqq Z(x) \subseteq G,$$

where the order of Z is at least p by Theorem 4.2. On the other hand, the orders of Z and $Z(x)$ divide $|G| = p^2$. From these, we can only conclude that $Z(x)$ and G have the same order, p^2; hence $Z(x) = G$. We have shown that $Z(x) = G$ for each $x \in G$, and this finishes the whole proof. □

We then ask whether a group of order p^3 is always abelian or not, and the answer is not:

EXAMPLE 4.2. *The dihedral group D_4 of order $8 = 2^3$ is not abelian.*

COROLLARY 4.5. *Let p be a prime number Then every group G of order p^2 is either one of the following two:*

(1) *a cyclic group of order p^2;*
(2) *a product of two cyclic groups of order p.*

PROOF. We first note that every non-identity element x of G has order dividing p^2. Let x be a non-identity element of G. If the order of x is p^2, then G is a cyclic of order p^2. Otherwise, every non-identity element of G has order p, so let x, y be two non-identity elements of G of order p, and let H_1, H_2 be the cyclic subgroups of order p generated by x and y, respectively. Because $p^2 > p$, we can always choose y to be not a power of x. We then try to apply Part (3) of Proposition 2.16 by verifying the required conditions: Since $y \notin H_1$, $H_1 \cap H_2$ is a proper subgroup of H_2; hence $H_1 \cap H_2 = \{1\}$. On the other hand, G is abelian by Proposition 4.3, so that H_1 and H_2 are normal and hence their product H_1H_2 is a subgroup of G by Part (2) of Proposition 2.16. Thus, its order is either p or p^2. But, again, since $y \notin H_1$, H_1H_2 properly contains H_1; hence the order of H_1H_2 must be p^2; hence $H_1H_2 = G$. Therefore, by Part (3) of Proposition 2.16 we conclude that G is isomorphic to $H_1 \times H_2$. □

REMARK 4.4. *We note that the number of possibilities for groups of order p^n increases rapidly with n.*

EXAMPLE 4.3. *There are 5 isomorphism classes of groups of order 2^3 among them 3 are abelian, and 14 classes of groups of order 2^4 among them 5 are abelian, etc. For more about these results, one may refer to Herstein's book,* Topics in Algebra, *2nd ed.*

Exercises for Chapter 4.1

EXERCISE 4.1. *Show that there are 3 conjugacy classes of conjugate elements in S_3.*

EXERCISE 4.2. *Show that there are 5 conjugacy classes of conjugate elements in S_4.*

EXERCISE 4.3. *Find the number of conjugates of the cycle (123) in S_4.*

EXERCISE 4.4. *Find the number of conjugates of $(12)(34)$ in S_4.*

EXERCISE 4.5. *Find the number of conjugates of $(12)(34)$ in A_4.*

EXERCISE 4.6. *If G/Z is cyclic, prove that G is abelian, i.e., $Z = G$.*

EXERCISE 4.7. *Prove that every non-abelian group of order p^3 has center of order p, where p is a prime number.*

2. The Alternating Group A_5 is Simple

REMARK 4.5. *The icosahedral group I plays an important role in Galois theory in connection with the problem of solving real quintic polynomial equations by radicals. The history tells us that a real polynomial equation is solvable by radicals if and only if the associated Galois group is solvable.*

The connection between the symmetries of the icosahedron and unsolvable real quintics is subtle (F. Klein, Lectures on the icosahedron, and the solution of equations of the fifth degree, *Trubner and Co.,* 1888*). It is* *based on the fact that a real quintic with Galois group* S_5 *is an example of a degree* 5 *equation for which no root formula exists. Any irreducible (over the rational numbers) quintic with exactly* 3 *real roots provides such an example, because in this case its Galois group is* S_5*. (Corollary* 6.10 *of the author's book, Computational Algebra.)*

*It was Klein (*1849 - 1925*) who launched the well-known* "Erlangen Program" *in* 1872*, the study of the properties of a space that are invariant under a given group of transformations, which profoundly influenced the evolution of later mathematics, perhaps until now.*

As a continuation of Example 3.20, we will again figure out all the elements of the icosahedral group G of rotational symmetries of order 60 of a dodecahedron, the group of orientation-preserving symmetries of a regular dodecahedron D,

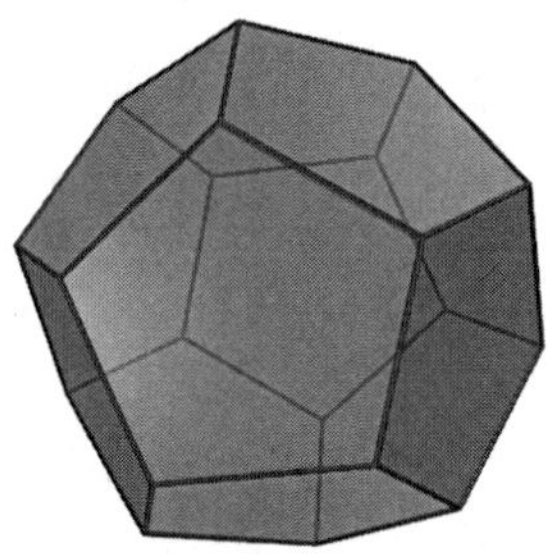

Dodecahedron

by *letting G act on itself by conjugation.* For this, we partition the group G according to the orders of its elements, and this will establish a numerical equation, (4.9):

$$60 = 1 + 15 + 20 + 24,$$

where the numbers on the right side are the number of elements of order $1, 2, 3, 5$, respectively.

But, unlike S_3 *in Example 4.1 in the last Section, the number* $24 = 2^3 \cdot 3$ *is not a divisor of* $60 = 2^2 \cdot 3 \cdot 5$*, the order of the group* G*, so that the above equation is not the Class Equation in (4.7) for the icosahedral group* G*.*

REMARK 4.6. *The problem is that elements of the same order do* not *always belong to the same conjugacy class.*

We shall see very soon that there are two conjugacy classes of elements of order 5 of the icosahedral group, so the number

$$24 = 12 + 12,$$

and this modification gives us the right Class Equation for I:

$$60 = 1 + 15 + 20 + 12 + 12.$$

2.1. The Class Equation for Icosahedral Group. Let G denote the icosahedral group I of rotational symmetries of order 60 of a dodecahedron. In Example 3.20, we have learned that its order is 60. Since all the symmetries of the regular dodecahedron D, which fixes the origin, are rotations, we need to examine these rotations in detail. We may assume that D are centered at the origin of $\mathbb{R}^3$. There are three different rotations of *prime* orders 5, 3, and 2, respectively:

(1) rotations by multiples of $2\pi/5$ about the centers of the face of D;
(2) rotations by multiples of $2\pi/3$ about the vertices;
(3) rotations by multiples of π about the centers of the edges.

From the geometric figure of D, we already know that D has 12 faces, 30 edges, and 20 vertices. Since what we want is information on the elements of the group G, we consider the stabilizer of vertices, edges, and faces:

(1) Each of the 12 faces has a stabilizer of order 5 and the opposite faces have the same stabilizer; hence, there are 6 subgroups of order 5 - the stabilizers of faces. Moreover, each cyclic subgroup of order 5 contains 4 elements of order 5, thus G contains $6 \times 4 = 24$ elements of order 5.
(2) Each of the 20 vertices has a stabilizer of order 3 and the opposite vertices have the same stabilizer; hence, there are 10 subgroups of order 3 - the stabilizers of vertices. Moreover, each cyclic

subgroup of order 3 contains two elements of order 3, thus G contains $10 \times 2 = 20$ elements of order 3.

(3) Each of the 30 edges has a stabilizer of order 2 and the opposite edges have the same stabilizer; hence there are 15 subgroups of order 2 - the stabilizers of edges. Thus G contains 15 elements of order 2.

(4) Finally, G has the identity element 1 and its order is 1.

Then by adding all these numbers according to orders, we see that

$$1 + 15 + 20 + 24 = 60, \tag{4.9}$$

thus we have counted all the elements of D. But, 24 never divides 60, so this equation is not the Class Equation for D. On the other hand, we know that every element in the same conjugacy class has the same order (Corollary 4.1), so by subdividing this partition of G based on the orders of elements further, we can obtain the right Counting formula for G.

We note that Proposition 3.3 says that *cyclic* subgroups of order 5, the stabilizers of the faces, are conjugate to each other, and similarly, *cyclic* subgroups of order 3, the stabilizers of the vertices, are conjugate to each other, and subgroups of order 2, the stabilizers of the edges, are conjugate to each other. Although this information does not tell us whether there is one conjugate class for a given element of a cyclic group of order r, this certainly forces us to consider G-action itself by conjugation.

So let G act on itself by conjugation. Then, in the case of those 15 non-identity elements of order 2, they all belong to the same conjugacy class because conjugation of one non-identity element of order 2 by $g \neq 1$ is again a non-identity element of order 2.

In the case of those 20 non-identity elements of those stabilizers of order 3 of vertices, if x is a non-identity element of order 3 then so is x^2. Since there is a rotation by which the vertex v is moved to *any* other vertex, say, w, and since $x \in G_v$, the stabilizer of v, by Part (2) of Proposition 3.3, x is conjugate to any counterclockwise rotation by $2\pi/3$

about w, i.e., the one in G_w. But, this does not tell us whether x and x^2 are conjugate to each other or not.

To examine this in a more detailed manner, let x be a counterclockwise rotation by $2\pi/3$ about a vertex v and x' a counterclockwise rotation by the same angle $2\pi/3$ about the opposite vertex v' of v. Then since there is a rotation which moves v to v', arguing as we have done above, we see that x and x' are conjugate to each other. Moreover, it is the geometry which tells us that the clockwise rotation, i.e., x^2, about v is exactly the same as the counterclockwise rotation x' about v'. Thus, x and $x^2 = x'$ are conjugate to each other. *Thus, those* 20 *non-identity elements of those stabilizers of order* 3 *of vertices all belong to the same conjugacy class,* $C_{2\pi/3}$.

Now, in the case of 24 non-identity elements of those stabilizers of order 5 of faces, a similar argument as we have just done above enables us to conclude that the 12 rotations by $2\pi/5$, $-2\pi/5$ are conjugate to each other and they belong to the same conjugacy class $C_{2\pi/5}$, and the 12 rotations by $2\cdot\ 2\pi/5$, $-2\cdot\ 2\pi/5$ are conjugate to each other and they belong to the same conjugacy class $C_{4\pi/5}$. But these two conjugacy classes can not be the same, for, if they were the same, then it would contain all the 24 elements, but is impossible because the number 24 never divides $|G| = 60$. *Thus, there are two different conjugacy classes of elements of order* 5, *each consisting of* 12 *elements of order* 5, and thus the Class Equation for D is:

$$1 + 15 + 20 + 12 + 12 = 60. \tag{4.10}$$

2.2. The Icosahedral Group I is Simple. We start our discussion with a definition:

DEFINITION 4.2. *A group G is called **simple** if it is not the trivial group and it does not have any proper normal subgroup.*

EXAMPLE 4.4. *Cyclic groups of prime order has no proper subgroups, so that they are simple.*

To prove that A_5 is simple, we are going to prove first that the icosahedral group I is simple using the Class Equation for D in (4.10), and this approach requires the following simple lemma:

LEMMA 4.1. *Let N be a normal subgroup of a group G. Then*

(1) *If N contains an element x, then it contains the conjugacy class C_x of x. Equivalently, we have*

$$N = \coprod_{x \in N} C_x. \tag{4.11}$$

(2) *The order of N is the sum of the orders of those conjugacy classes which it contains.*

PROOF. Once Part (1) is proved, Part (2) holds automatically, and the proof of Part (1) is very easy. □

In fact, we have:

COROLLARY 4.6. *A subgroup N of a group G is normal in G if and only if*

$$N = \coprod_{x \in N} C_x.$$

PROOF. Exercise. □

REMARK 4.7. *One more equivalent statement for a subgroup H of a group G being normal in G is given in Corollary 4.10 below.*

THEOREM 4.3. *The icosahedral group I is simple.*

PROOF. (By contradiction) Suppose that I is not simple, say N is its non-trivial proper normal subgroup. Then the order of N would be a proper divisor of $60 = 2^2 \cdot 3 \cdot 5$, and the possible non-trivial proper divisors of 60 are:

$$\begin{aligned} 2^2 \cdot 3 &= 12,\ 2^2 \cdot 5 = 20,\ 2^2 = 4; \\ 2 \cdot 3 \cdot 5 &= 30,\ 2 \cdot 3 = 6,\ 2 \cdot 5 = 10,\ 2; \\ 3 \cdot 5 &= 15,\ 3,\ 5. \end{aligned}$$

On the other hand, Part (2) of Lemma 4.1 tells us particularly that a possible order of N must be also the sum of some of those numbers in the left side of the Class Equation in (4.10), *including the number* 1. But, we can easily verify that there is no such sum. Therefore, I must be simple. □

2.3. The Alternating Group A_5 is Simple. We start with a picture:

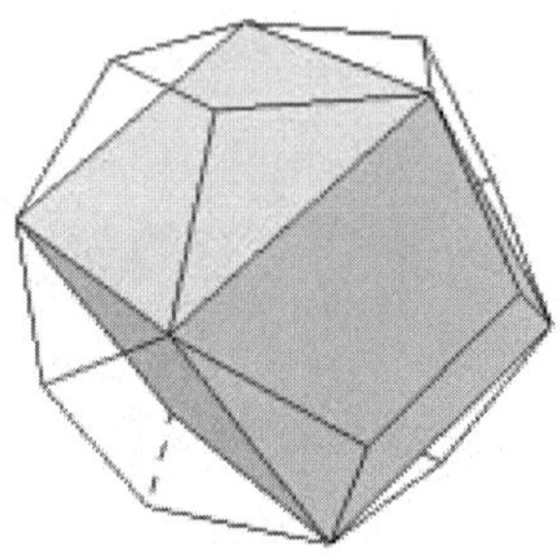

The above figure shows one of the five cubes inscribed in a dodecahedron.

THEOREM 4.4. *The icosahedral group I is isomorphic to the alternating group A_5.*

PROOF. We first note that there are in total 5 cubes inscribed in the dodecahedron, and to these 5 figures the group I acts, and hence by Proposition 3.6, we have the associated permutation representation $\varphi : I \longrightarrow S_5$. Consider the kernel of φ, a normal subgroup of I. Since I is simple by Theorem 4.3, $\ker \varphi$ is either $\{1\}$ or I. If $\ker \varphi$ were I, then it would mean that the action of I on the set of 5 cubes were the trivial operation, but it is not the case. Thus, $\ker \varphi = \{1\}$; hence φ is injective. Thus, I is isomorphic to its image $\varphi(I)$ in S_5.

On the other hand, we have the sign homomorphism

$$sign : S_5 \longrightarrow \{\pm 1\}$$

which is surjective and its kernel is the alternating group A_5, consisting of all even permutations. We note that

$$|A_5| = |S_5|/2 = 5!/2 = 60,$$

by Corollary 2.11. Then the restriction of the sign homomorphism *sign* gives us another group homomorphism,

$$\sigma : \varphi(I) \longrightarrow \{\pm 1\}.$$

Now, if we can prove that the group homomorphism σ is a trivial homomorphism, i.e., the one sending everything in $\varphi(I)$ to 1, then this just means that $\varphi(I)$ is contained in the kernel A_5 of $sign : S_5 \longrightarrow \{\pm 1\}$. Then by size comparison, we can conclude that I and A_5 are isomorphic, because they have the same order 60. Thus, we are only left to prove that σ is not surjective:

If σ were surjective, then again by Corollary 2.11, we would have:

$$60 = |\varphi(I)| = |\ker(\sigma)| \cdot |\{\pm 1\}| = 2 \cdot |\ker(\sigma)|,$$

and hence $\ker(\sigma)$ would be a normal subgroup of order 30 of the group $\varphi(I) \simeq I$, but this is impossible because we have already proven in Theorem 4.3 that I is simple. □

COROLLARY 4.7. *A_5 is simple.*

In order for reader's convenience with terminologies used in other books, we leave a closing example of this Section:

EXAMPLE 4.5. *S_5 is not "solvable", for since A_5 is simple the series*

$$\{1\} \lhd A_5 \lhd S_5$$

is a "composition series", where $A_5/\{1\} = A_5$ is not an abelian group.

In fact, one can prove that A_5 is the smallest group that is not solvable.

Exercises for Chapter 4.2

EXERCISE 4.8. *Find the number of conjugates of the cycle* (123) *in S_5.*

EXERCISE 4.9. *Prove that no group of order* p^e, *where* p *prime and* $e > 1$, *is simple.*

EXERCISE 4.10. *Show that* A_3 *is the only proper normal subgroup of* S_3.

EXERCISE 4.11. *Prove that* A_5 *is the only non-trivial normal subgroup of* S_5.

EXERCISE 4.12. *Prove that the icosahedral group has no subgroup of order* 30.

3. The Induced Group Actions on Subsets

Suppose that we have a group action G on a set S. Then it induces another action of G on the set of subsets U of S in a natural manner by defining its action as follows:

$$g, U \longmapsto gU := \{gu : u \in U\}.$$

Here, gU is another subset of S, and it certainly satisfies the two requirements for a group action.

DEFINITION 4.3. *We call this action, the* ***induced group action*** *of* G *on the subsets of* S.

Since multiplication by an element g of G is a permutation of S, the order of U and that of gU must be the same. As usual, the stabilizer $Stab(U)$ of a subset U is the set of group elements g such that $gU = U$.

REMARK 4.8. *We note that* $gU = U$ *does not mean that each element of* U *is fixed under the action* g *of* G. *Rather, it means that the whole subset* U *is fixed by* g, *that is,* $gu \in U$ *whenever* $u \in U$.

EXAMPLE 4.6. *Let* O *be the octahedral group of* $24 = 1 + 3 \cdot 2 + 4 \cdot 2 + (3 + 6)$ *rotations of a cube, and* S *the set of* 8 *vertices of the cube (cf. Theorem 3.2). We in particular consider the* O *action on those subsets of order* 2, *that is, on unordered pairs of vertices. There are* ${}_7C_1 \cdot 8/2$ *or*

$$7 + 6 + 5 + 4 + 3 + 2 + 1 = 28$$

such pairs, and they decompose into three orbits for the group O:

(1) $\{$*pairs of vertices on an edge*$\}$;

(2) $\{$*pairs of opposite vertices on a face of the cube*$\}$;

(3) $\{$*pairs of opposite vertices on the cube*$\}$.

These orbits have orders, $12, 12,$ *and* 4, *respectively:*

$$28 = 12 + 12 + 4.$$

Then, the stabilizer of a pair of opposite vertices on a face contains two elements, the identity and the rotation by π about the face. Thus, we have:

$$|G| = 24 = 2 \cdot 12,$$

which agrees with the counting formula in Proposition 3.4.

PROPOSITION 4.4. *Let G act on a set S, and U a subset of S. Then G stabilizes U if and only if U is a union of G-orbits of elements of U.*

PROOF. Since the G-orbit of an element $u \in U$ is, by definition, the set of all those elements gu for $g \in G$, if G stabilizes a subset U, then U contains all the G-orbits of any element u of U. This proves the $\Longrightarrow$ direction. The converse is obvious. □

In this Section, we consider the induced G-action on subsets U of a group G when the group action of G on itself is left multiplication or conjugation.

3.1. G-action by left multiplication on subsets of a group G . Let a group G act on itself by left multiplication.

If H is a subgroup of G, then its orbits consist of the left cosets of H, namely, the coset space of H, and this action of G on cosets has been already studied in the previous Chapter.

Otherwise, we still have that any subset U of S has an orbit:

EXAMPLE 4.7. *Let $G = D_3$ be the dihedral group of symmetries of an equilateral triangle in the plane:*

$$G = \{x^i y^j : 0 \le i \le 2, 0 \le j \le 1,\ x^3 = 1, y^2 = 1, yx = x^2 y\},$$

and we let G act on itself by left multiplication. We know that it is the smallest non-abelian group (Corollary 2.2). Since ${}_5C_1 \cdot 6/2 = 15$ or

$$5+4+3+2+1=15,$$

there are in total 15 number of subsets of order 2, namely, unordered pairs of elements of G. We then decompose the set of these 15 subsets into 3+1 orbits for left multiplication as follows:

From the multiplication table in (2.2), we easily see that G has three subgroups of order 2,

$$H_1 = \{1, y\}, H_2 = \{1, xy\}, H_3 = \{1, x^2y\}.$$

Then their left cosets form three orbits of order 3,

$$\begin{aligned} \{H_1 &= \{1, y\}, xH_1 = \{x, xy\}, x^2H_1 = \{x^2, x^2y\}\}, \\ \{H_2 &= \{1, xy\}, xH_2 = \{x, x^2y\}, x^2H_2 = \{x^2, y\}, \\ \{H_3 &= \{1, x^2y\}, xH_3 = \{x, y\}, x^2H_3 = \{x^2, xy\}\}. \end{aligned}$$

So far, we have consumed 9 subsets of order 2. The other 6 subsets of order 2 clearly form a single orbit:

$$\{\{1, x\}, \{x, x^2\}, \{1, x^2\}, \{y, x^2y\}, \{xy, y\}, \{x^2y, xy\}\}.$$

Thus, we have:

$$15 = 3+3+3+6.$$

PROPOSITION 4.5. *Let a group G act on itself by left multiplication, U a subset of G. Then the order of the stabilizer $Stab(U)$ of U for the action of left multiplication divides the order of U.*

PROOF. By Proposition 4.4 with $G = Stab(U)$ there, we know that U is a union of $Stab(U)$-orbits for the action of the subgroup $Stab(U)$ on G. On the other hand, since the $Stab(U)$-orbit of an element $g \in G$ is, by definition, the set of all those elements hg for $h \in Stab(U), g \in G$, these $Stab(U)$-orbits are right cosets $Stab(U)g$. Thus, U is a union of right cosets; consequently, the order of U is a multiple of $|Stab(U)|$. □

REMARK 4.9. *This proposition is crucially applied in the proof of the First Sylow Theorem, Theorem 4.6.*

COROLLARY 4.8. *If the order of a group G and the order of its subset U are relatively prime, then $Stab(U) = \{1\}$.*

PROOF. This is because the order of the subgroup $Stab(U)$ of G divides both $|G|$ and $|U|$. □

3.2. G-action by conjugation on subsets of a group G . Before we examine several examples, we have an easy lemma:

LEMMA 4.2. *The dihedral group D_n has a normal subgroup of cyclic of order n.*

PROOF. Let D_n be the dihedral group generated by x, y with the three relations:

$$x^n = 1, y^2 = 1, yxy^{-1} = x^{-1}.$$

Let H be the cyclic subgroup of order n generated by x. Then since

$$yx^iy^{-1} = x^{-i},$$

for each $1 \leq i < n$, H is normal in D_n. Or, since $[D_n : H] = 2$, H is normal in D_n by Corollary 2.12. □

COROLLARY 4.9. *D_n is not simple.* □

EXAMPLE 4.8. *Let $G = D_3 \simeq S_3$ be the dihedral group of symmetries of an equilateral triangle in the plane:*

$$G = \{x^iy^j : 0 \leq i \leq 2, 0 \leq j \leq 1,\ x^3 = 1, y^2 = 1, yx = x^2y\},$$

and we let G act on itself by conjugation. Like we have done in Example 4.7, we can partition the 15 *subsets of order* 2 *into orbits, this time, for conjugation as follows:*

The subgroups H_1, H_2, H_3 form an orbit for conjugation, because

$$xH_1x^{-1} = H_3, x^2H_1x^{-2} = H_2.$$

The set $\{x, x^2\}$ clearly forms an orbit by itself. Then one easily checks that other orbits are,

$$\{\{1,x\},\{1,x^2\}\},$$
$$\{\{y,x^2y\},\{xy,y\},\{x^2y,xy\}\},$$
$$\{\{x,xy\},\{x^2,x^2y\},\{x^2,y\},\{x^2,xy\},\{x,x^2y\},\{x,y\}\}.$$

Thus, we again have

$$15 = 3 + 1 + 2 + 3 + 6.$$

For our later usage, we now consider the orbit under conjugation of a subgroup H of G. This orbit is the set S of conjugate subgroups:

$$S := \{gHg^{-1} : g \in G\}.$$

The set S clearly contains H, and G acts transitively on S by conjugation. We recall that a subgroup H of G is normal in G if and only if $gHg^{-1} = H$ for each $g \in G$ (Proposition 2.10). Thus,

COROLLARY 4.10. *A subgroup H of G is normal if and only if its orbit under conjugation consists of H alone.* □

REMARK 4.10. *One more equivalent statement for H being normal in G is given in Corollary 4.11 below.*

DEFINITION 4.4. *Let a group G act on itself by conjugation. The stabilizer of a subgroup H for the action of conjugation is called the* ***normalizer*** *of H and is denoted by*

$$N(H) = \{g \in G : gHg^{-1} = H\}.$$

Thus, the normalizer $N(H)$ always contains the subgroup H, because $hHh^{-1} = H$ when $h \in H$. Thus, by Lagrange's Theorem (Corollary 2.8), $|H|$ divides $|N(H)|$, and $|N(H)|$ divides $|G|$.

Moreover, H is automatically normal in $N(H)$ by definition of $N(H)$. We also note that $N(H)$ is the largest group containing H as a normal subgroup. In particular, we have:

COROLLARY 4.11. *$N(H) = G$ if and only if H is a normal subgroup of G.* □

In summary, we have seen that the normalizer $N(H)$ of the subgroup H is nothing but the "centralizer" of H under the action of conjugation on the set S on which G clearly acts transitively. Thus, the Counting Formula in Proposition 3.4 reads:

$$|G| = |N(H)| \cdot |\{\text{conjugate subgroups of } H\}|, \tag{4.12}$$

and hence the number of conjugate subgroups of H is equal to the index

$$[G : N(H)].$$

REMARK 4.11. *We need (4.12) in the proof of the third Sylow Theorem, Theorem 4.8.*

EXAMPLE 4.9. *In Example 4.8 with $G = D_3$, the three subgroups H_i are conjugate to each other because they form an orbit, and hence $|N(H_i)| = 2$ by (4.12); consequently, $N(H_i) = H_i$ for each $i = 1, 2, 3$. This proves by Corollary 4.11 that each of the three subgroups H_i of order 2 is not normal in D_3.*

REMARK 4.12. *Before we move into the next section which is mainly for the proofs of the Sylow Theorems, we provide in what follows in this Section several examples of conjugation action of a group G on itself.*

EXAMPLE 4.10. *Let $G = S_3 \simeq D_3$ act on itself by conjugation. It consists of $3! = 3 \cdot 2 \cdot 1 = 6$ elements, and this group consists of:*

(1) 3 *transpositions:*

$$C_{(12)} = \{(12), (13), (23)\},$$

(2) 2 *3-cycles:*

$$C_{(123)} = \{(123), (132)\},$$

(3) *the identity:*

$$C_{(1)} = \{1\}.$$

Then, one can easily check that each of these 3 *disjoint collections forms a conjugacy class, and hence the Class Equation in (4.7) for* S_3 *reads:*

$$1 + 3 + 2 = 6.$$

Moreover, one can also easily verify that:

$$A_3 = C_{(1)} \cup C_{(123)} \simeq \mathbb{Z}_3,$$

which is abelian, and normal subgroups of S_3 *are:*

$$\{1\}, A_3, S_3.$$

Thus, $S_3 \simeq D_3$ *is not simple, and* A_3 *is simple.*

We also note that the number of conjugate classes is $3 = p(3)$*, where* $p(n)$ *is the partition function defined in the Subsection, Additive Factorization Theory?, of Chapter 6 of this book.*

EXAMPLE 4.11. *Let* $G = S_4$ *act on itself by conjugation. It consists of* $4! = 4 \cdot 3 \cdot 2 \cdot 1 = 24$ *elements, and this group consists of:*

(1) 6 *transpositions:*

$$C_{(12)} = \{(12), (13), (14), (23), (24), (34\},$$

(2) 8 3-*cycles:*

$$C_{(123)} = \{(123), (132); (124), (142); (134), (143); (234), (243)\},$$

(3) 3 *products of two disjoint transpositions:*

$$C_{(12)(34)} = \{(12)(34), (13)(24), (14)(23)\},$$

(4) 6 4-*cycles:*

$$C_{(1234)} = \{(1234), (1243); (1324), (1342); (1423), (1432)\}.$$

(5) *the identity:*

$$C_{(1)} = \{1\}.$$

Then one can easily check that each of these 5 *disjoint collections forms a conjugacy class, and hence the Class Equation in (4.7) for* S_4 *reads:*

$$1 + 3 + 6 + 6 + 8 = 24.$$

Then since $(1 + 3)$ *divides* 24 *and* $(1 + 3 + 8)$ *didives* 24*, we can have a normal subgroup of order* 4 *and/or a normal subgroup of order* 12*. On the other hand, one easily verifies that*

$$\begin{aligned} A_4 &= C_{(1)} \cup C_{(123)} \cup C_{(12)(34)}, \\ V &= C_{(1)} \cup C_{(12)(34)}, \end{aligned}$$

are both normal subgroups of S_4*, where* V *is the Klein* 4 *group. Thus normal subgroups of* S_4 *are:*

$$\{1\}, V, A_4, S_4.$$

Therefore, neither S_4 *nor* A_4 *is simple.*

We also note that the number of conjugate classes is $5 = p(4)$*, where* $p(n)$ *is the partition function.*

EXAMPLE 4.12. *Let* $G = A_4$ *act on itself by conjugation. It consists of* $4!/2 = 12$ *elements, and one can verify that this group consists of:*

(1) 4 3-*cycles:*

$$C_{(123)} = \{(123), (134), (142), (243)\},$$

(2) *another* 4 3-*cycles:*

$$C_{(132)} = \{(132), (143), (124), (234)\},$$

(3) 3 *products of two disjoint transpositions:*

$$C_{(12)(34)} = \{(12)(34), (13)(24), (14)(23)\},$$

(4) *the identity*

$$C_{(1)} = \{1\}.$$

Then one can verify that each of these 4 *disjoint collections forms a conjugacy class, and hence the Class Equation in (4.7) for* A_4 *reads:*

$$1 + 4 + 4 + 3 = 12.$$

Then since $(1+3)$ *divides* 12, *we can have a normal subgroup of order* 4. *On the other hand, one easily verifies that the Klein* 4 *group*

$$V = C_{(1)} \cup C_{(12)(34)}$$

is a normal subgroup of A_4*; hence* A_4 *is not simple. [Cf. the result of Example 4.11 above].*

REMARK 4.13. *In particular, this example shows that* A_4 *does not have an element of order* 6.

COROLLARY 4.12. A_4 *can not have a subgroup of order* 6.

PROOF. By applying the same arguments as the one in the proof of Theorem 4.3, we can prove this result, and it goes as follows: If it would have a subgroup H of order 6, then it would have index $[A_4 : H] = 2$ by Corollary 2.12; hence it would be a normal subgroup of A_4; hence we would be able to express its order as a sum of orders of those conjugacy classes in (4.11). But, if so, then we *cannot* have the "right" Class Equation for A_4, because 6 is never equal to a sum of these numbers, $1, 4, 4, 3$ in Example 4.12. □

REMARK 4.14. *Thus, the converse of Lagrange's Theorem (Corollary 2.8) does not hold in general.*

EXAMPLE 4.13. *Let* $G = S_5$ *act on itself by conjugation. We know that it has* $5! = 120$ *elements, and one can verify that this group consists of:*

(1) $C_{(12)} = \{10$ *transpositions*$\}$;
(2) $C_{(123)} = \{20$ 3-*cycles*$\}$;
(3) $C_{(12)(34)} = \{15$ *products of two disjoint transpositions*$\}$;
(4) $C_{(1234)} = \{30$ 4-*cycles*$\}$;

(5) $C_{(12)(345)}$ = {20 *products of two disjoint transpositions and 3-cycles*};
(6) $C_{(12345)} = \{24$ 5-*cylces*$\}$;
(7) $C_{(1)} = \{1\}$.

Then one can verify that each of these 7 *disjoint collections forms a conjugacy class, and hence the Class Equation in (4.7) for* S_5 *reads:*

$$1 + 10 + 20 + 15 + 30 + 20 + 24 = 120.$$

We also note that the number of conjugate classes is $7 = p(5)$, *where* $p(n)$ *is the partition function.*

Example 4.14. *Let* $G = A_5$ *act on itself by conjugation. It consists of* $5!/2 = 60$ *elements, and one can verify that this group consists of:*

(1) $C'_{(123)} = \{20$ 3-*cycles*$\}$;
(2) $C'_{(12)(34)} = \{15$ *products of two disjoint transpositions*$\}$;
(3) $C'_{(12345)} = \{12$ 5-*cycles*$\}$ *containing* (12345);
(4) $C'_{(12354)} = \{12$ *another* 5-*cycles*$\}$ *containing* (12354);
(5) *the identity* $\{1\}$.

Then one can verify that each of these 5 *disjoint collections forms a conjugacy class, and hence the Class Equation in (4.7) for* A_5 *reads:*

$$1 + 20 + 15 + 12 + 12 = 60.$$

Moreover, by the same argument as the one in the proof of Theorem 4.3, we see that A_5 *is simple.*

On the other hand, the following theorem (Theorem 4.5) provides us with an affirmative answer to the question, whether A_n is simple or not. To establish Theorem 4.5 below, we need two lemmas:

Lemma 4.3. A_n *with* $n \geq 3$ *is generated by* 3-*cycles.*

PROOF. Since every element of A_n is a product of the product of transpositions of the form,

$$\begin{aligned}(12)(34) &= (142)(143),\\ (12)(13) &= (132),\ \textit{or}\\ (12)(12) &= 1,\end{aligned}$$

we are done. □

LEMMA 4.4. *If a normal subgroup N of A_n with $n \geq 3$ contains a 3-cycle (abc), then $N = A_n$.*

PROOF. When $n = 3$, we obviously know that (123) is in N. Then $(132) = (123)(123) \in N$; hence $N = A_3$.

When $n \geq 4$, we may assume that $(123) \in N$. We then note that there are in total 5 different forms of 3-cycles:

$$(12a), (1a2), (1ab), (2ab), (abc)$$

where a, b, c are distinct and ≥ 3. On the other hand, since N is normal in A_n we have, for any $a \geq 4$,

$$(12a) = ((12)(3a))(123)(123)((3a)(12))^{-1} \in N.$$

In addition to this, for $a, b, c \geq 3$,we have

$$\begin{aligned}(1a2) &= (12a)(12a),\\ (1ab) &= (12b)(12a)(12a),\\ (2ab) &= (12b)(12b)(12a),\\ (abc) &= (12a)(12a)(12c)(12b)(12b)(12a),\end{aligned}$$

and hence they are all in N. Therefore, N contains all 3-cycles and hence $N = A_3$. □

COROLLARY 4.13. *A_n with $n \geq 3$ is generated by those 3-cycles (rsk),where r and s are two distinct numbers $\leq n, k \neq r, s$, and k runs through from 1 to n.* □

THEOREM 4.5. *The alternating group A_n is simple if and only if $n \neq 4$.*

PROOF. Since we have already verified in Example 4.10 and Example 4.11 that A_3 is simple but A_4 is not simple, we only need to prove that if $n \geq 5$, then A_n is simple. For this, let N be a non-trivial normal subgroup of A_n. If we *can* prove that $N = A_n$, by showing that a 3-cycle $(rsk) \in N$, where r and s are two distinct numbers $\leq n, k \neq r, s$, and k runs through from 1 to n, then we are done by Corollary 4.13, and it goes as follows: Since N contains a non-identity even permutation x, we then note that 'this' x may be one of the following forms of the products of *disjoint* cycles:

$$\begin{aligned} x &= (123), \\ x &= (123)\sigma, \\ x &= (12...r)\sigma, \text{ with } r \geq 4, \\ x &= (123)(456)\sigma, \text{ or} \\ x &= (12)(34)\sigma, \end{aligned}$$

where σ is a product of disjoint 2-cycles, *none of which involves the letters, respectively in each case, in* $(12...r), (123)(456), (123),$ *and* $(12)(34)$.Under this setting, we have:

(1) If $x = (123) \in N$, then by Lemma 4.4 we are done.
(2) If $x = (123)\sigma \in N$, then $x^2 = (132) \in N$; hence we are done by Lemma 4.4.
(3) If $x = (12...r)\sigma \in N$ with $r \geq 4$, then for a particular $y = (123)$ $\in A_n$ that is automatically *disjoint* from σ by our choice of σ, we see that $x^{-1}(yxy^{-1})$ is contained in N, but we also have:

$$\begin{aligned} x^{-1}(yxy^{-1}) &= (1r...32)\sigma^{-1}(123)(12...r)\sigma(123)^{-1} \\ &= \sigma^{-1}\sigma(1r...32)(123)(12...r)(132) \\ &= (13r), \end{aligned}$$

and hence we have shown that $(13r) \in N$. Thus, by Corollary 4.13, we are done.

(4) If $x = (123)(456)\sigma \in N$, then for a particular $y = (124) \in A_n$ that is automatically *disjoint* from σ by our choice of σ, we see that $x^{-1}(yxy^{-1})$ is contained in N, but we also have:

$$\begin{aligned} x^{-1}(yxy^{-1}) &= (132)(465)\sigma^{-1}(124)(123)(456)\sigma(142) \\ &= \sigma^{-1}\sigma(132)(465)(124)(123)(456)(142) \\ &= (14263), \end{aligned}$$

and hence $(14263) \in N$. Thus, by Part (3), we are done.

(5) If $x = (12)(34)\sigma \in N$, then for a particular $y = (123) \in A_n$ that is automatically *disjoint* from σ by our choice of σ, we see that $x^{-1}(yxy^{-1})$ is contained in N, and we also have:

$$\begin{aligned} x^{-1}(yxy^{-1}) &= (12)(34)\sigma^{-1}(123)(12)(34)\sigma(132) \\ &= \sigma^{-1}\sigma(12)(34)(123)(12)(34)(132) \\ &= (13)(24), \end{aligned}$$

so that $z = (13)(24) \in N$. Then moreover since $n \geq 5$, there exists a number $k \leq n$ that is distinct from $1, 2, 3,$ and 4 and we let $u = (13k) \in A_n$ be a particular even permutation. Then since $z = (13)(24) \in N$, uzu^{-1} is also an element of N; hence $z(uzu^{-1}) \in N$, but we also have:

$$\begin{aligned} z(uzu^{-1}) &= (13)(24)(13k)(13)(24)(1k3) \\ &= (13k), \end{aligned}$$

and hence we have shown that $(13k) \in N$. Thus, by Corollary 4.13, we are done.

□

Exercises for Chapter 4.3

EXERCISE 4.13. *Verify the details of Example 4.10.*

EXERCISE 4.14. *Verify the details of Example 4.11.*

EXERCISE 4.15. *Verify the details of Example 4.13.*

EXERCISE 4.16. *Verify the details of Example 4.12.*

EXERCISE 4.17. *Verify the details of Example 4.14.*

EXERCISE 4.18. *Let H, K be subgroups of a group G such that $H \subseteq K$. Prove that H is normal in K if and only if $K \subseteq N(H)$.*

EXERCISE 4.19. *Determine the orbits for the action of D_3 on S by conjugation, where S is the set of subsets of order 2 of D_3.*

EXERCISE 4.20. *Determine the orbits for the action of D_3 on S by conjugation, where S is the set of subsets of order 3 of D_3.*

EXERCISE 4.21. *Determine all subgroups of D_4, and divide them into conjugacy classes.*

EXERCISE 4.22. *Determine the Class Equation of each of the following groups: (i) the quaternion, (ii) the Klein 4 group, (iii) D_5, (iv) D_6, (v) D_n.*

EXERCISE 4.23. *Determine the Class Equation for the octahedral group O. Find and prove that there are only two proper normal subgroups of O.*

EXERCISE 4.24. *Show that a product of non-necessarily disjoint cycles is even if and only if it contains an even number of cycles of even length.*

EXERCISE 4.25. *Prove that every permutation of order 14 in S_{10} is odd.*

EXERCISE 4.26. *Prove that S_n is generated by $n-1$ transpositions.*

EXERCISE 4.27. *Let N be a subgroup of order 5 of a group of odd order. Prove that N is contained in the center of G.*

4. The Sylow Theorems

Let G be a group of order n, and p a prime number diving n. Then

$$n = p^e m$$

for some number m not divisible by p.

DEFINITION 4.5. *A group G is called a p-**group** if its order is a power of a prime number p.*

LEMMA 4.5. *The number N of subsets of order p^e in a set of $n = p^e m$ elements with $(p, m) = 1$ is the binomial coefficient*

$$N = \binom{n}{p^e} = \frac{n}{p^e} \cdot \frac{(n-1)}{(p^e-1)} \cdots \frac{(n-k)}{(p^e-k)} \cdots \frac{(n-p^e+1)}{1},$$

and moreover N is not divisible by p.

PROOF. It is a standard fact in Combinatorics that N is equal to the binomial coefficient. To show that p does not divide N, we write $k = p^j q$ with $(p, q) = 1$ and $j < e$. Then $n - k = p^j(p^{e-j}m - q)$ and $p^e - k = p^j(p^{e-j} - q)$. Thus whenever some p^j divides the numerator, it is cancelled out by the same p^j in the denominator; consequently the right side is not divisible by p. □

THEOREM 4.6. *(First Sylow Theorem) Let G be a group of order $p^e m$ with $(p, m) = 1$. Then there exists a subgroup of order p^e.*

PROOF. Let G act on itself by left multiplication and let S be the collection of all subsets of of G of order p^e. Then the group G also acts on S by left multiplication; hence the elementary counting formula in (3.6) reads

$$N = |S| = \sum_{orbits\ \mathcal{O}} |\mathcal{O}|$$

Since N is not divisible by p by Lemma 4.5, at least the order of one orbit $\mathcal{O}$ is not divisible by p. Let this orbit be the orbit of a set U of order p^e, and then the stabilizer $Stab(U)$ of U is a subgroup of G and by Proposition 4.5 its order divides the order p^e of U, and hence $|Stab(U)|$ is a power of p. On the other hand, by Proposition 3.4 we have the Counting Formula:

$$p^e m = |G| = |Stab(U)| \cdot |\mathcal{O}| .$$

Then since we have chosen the orbit $\mathcal{O}$ such that its order does not divides p^e, we must have:

$$p^e = |Stab(U)|,$$

and this proves our theorem. □

REMARK 4.15. *The above proof is quite remarkable in the sense that it is not based on an induction on either $|G|$ or the exponent e of p^e, where $|G| = p^e m$ with $(p, m) = 1$.*

EXAMPLE 4.15. *A group of order $6 = 2 \cdot 3$, for instance $S_3 \simeq D_3$, always has an element of order 2 and an element of order 3.*

DEFINITION 4.6. *Let G be a group of order $p^e m$ with $(p, m) = 1$ and prime p. The subgroups of order p^e whose existence is guaranteed by Theorem 4.6, are called **Sylow p-subgroups** or simply **Sylow-subgroups**.*

Thus, a Sylow p-subgroup of G is a p-subgroup whose index m in G is not divisible by p. Sylow's two remaining theorems tell us more information about them.

COROLLARY 4.14. *(Cauchy's Theorem) If a prime p divides the order of a group G, then G contains an element of order p.*

PROOF. Let G be a group of order $p^e m$ with $(p, m) = 1$. By Theorem 4.6, G has a Sylow p-subgroup H of order p^e. Let $x \neq 1$ be an element of H. Then the order of x divides p^e, so it is p^r for some number r in $0 < r \leq e$. Then $x^{p^{r-1}}$ clearly has order p. □

REMARK 4.16. *We note that Cauchy's Theorem is not obvious at all unless the First Sylow Theorem is established in advance. But, historically, it was not that case. It is well-known that Cauchy (1789 - 1857) established his theorem (Corollary 4.14) before Sylow proved his First and other two Theorems in 1872. Thus, the First Sylow Theorem is in some sense a generalization of Cauchy's Theorem. On the other hand, one can prove the First Sylow Theorem using Cauchy's Theorem, and hence we may say that they are equivalent.*

COROLLARY 4.15. *A group G is a p-group if and only if the order of each element of G is a power of p.*

PROOF. The $\Longleftarrow$ direction: Let $p^e m$ with $(p, m) = 1$ be the order of G. If $m \neq 1$, m is divisible by some prime q other than p, and hence by Cauchy's Theorem (Corollary 4.14)), G has an element of order q, but this is impossible. The $\Longrightarrow$ direction holds by Lagrange's Theorem (Corollary 2.8). □

THEOREM 4.7. *(Second Sylow Theorem) Let K be a subgroup of a group G whose order is divisible by a prime number p, H a Sylow p-subgroup of G. Then there exists a conjugate subgroup $H' = aHa^{-1}$ such that $K \cap H'$ is a Sylow p-subgroup of K.*

PROOF. We first recall the followings: By Sylow's first theorem, we know that G has H. Let G act on itself by left multiplication and S be the coset space G/H of H. Then by Lemma 3.3 G acts transitively on S and H is the stabilizer of itself, $1H = H$; then by Part (2) of Proposition 3.3 the stabilizer of *any* coset aH in the coset space is aHa^{-1}.

Now, since we are looking for a Sylow p-subgroup of K, we restrict the action of G on S to K, and by doing so, we decompose the set S into a union of disjoint K-orbits. We then note that, since H is a Sylow p-group, the order of S is prime to p by (3.4); hence by the elementary counting formula in (3.6), the order of at least one K-orbit, say $\mathcal{O}$, must be prime to p. Let this $\mathcal{O}$ be the K-orbit of aH for some $a \in G$ so that the stabilizer of aH under the group action G on S is $H' = aHa^{-1}$. *Then the stabilizer of aH under the induced restricted action of K on S is definitely* $H' \cap K$ as we have seen in the proof of Proposition 3.5; hence the index $[K : H' \cap K] = |\mathcal{O}|$, so it is prime to p. But, on the other hand, since H' is a conjugate of H, it is also a p-group. (This is because H and H' are isomorphic.) Thus its subgroup $H' \cap K$ is also a p-group, and we have already shown above that its index $[K : H' \cap K] = |\mathcal{O}|$ is prime to p. Therefore, by the counting formula in Lemma 2.27 $H' \cap K$ is indeed a Sylow p-subgroup of K. □

COROLLARY 4.16. (1) *If K is a subgroup of a group G such that K itself is a p-group, then K is contained in a Sylow p-subgroup of G.*

(2) *The Sylow p-subgroups of a group G are all conjugate.*

PROOF. Part (1): Since K is a p-group, it is its own Sylow p-subgroup. Let H be a Sylow p-subgroup of G. Then by Sylow's second theorem we know that there exists a conjugate $H' = aHa^{-1}$ of H such that $H' \cap K$ is a Sylow p-subgroup of K. Thus, $K = H' \cap K \subseteq H'$, where H' is also a Sylow p-subgroup of G because it is a conjugate of H.

Part (2): Let K and H be two Sylow p-subgroups of G. Then by the proof of Part (1), there exists a conjugate H' of H which contains K. Then since they have the same order, $K = H'$, a conjugate of H. □

THEOREM 4.8. *(Third Sylow Theorem) Let G be a group of order $n = p^e m$ with $(p, m) = 1$ and p a prime number. Let s be the number of the set S of all Sylow p-subgroups of G. Then*

(1) *s divides m;*
(2) *$s \equiv 1 \pmod{p}$.*

PROOF. By Corollary 4.16, the Sylow p-subgroups of G are all conjugate to a given one, say H. Let

$$S = \{H = H_1, H_2, ..., H_s\}$$

be the set of all conjugates of H, so that $s = |S|$. Under this setting we need to prove Part (1) and Part (2).

Part (1): Let G act on itself by conjugation. Let $N(H)$ be the normalizer, i.e., the stabilizer, of H in G. Then $s = [G : N(H)]$ by (4.12).

Then since H is a subgroup of $N(H)$, we have

$$|N(H)| = |H| \cdot m_1 = p^e m_1,$$

for some natural number $m_1 \geq 1$, but $|N(H)|$ should divide $|G| = p^e m$ by Lagrange's Theorem (Corollary 2.8) because $N(H)$ is a subgroup of

G. Thus, m_1 divides m; and this forces us to have an equality:

$$p^e m_1 \cdot s = |N(H)| \cdot s = |G| = p^e m_1 \cdot \frac{m}{m_1}.$$

Thus, $s = m/m_1$, that is, s is a divisor of m.

Part (2): Let H act on the set S by conjugation. Then S is decomposed into a union of disjoint H-orbits for the action of $H = H_1$ by conjugation on S. We then easily see that the orbit of H consists of H alone. On the other hand, by Proposition 3.4, we know that the order of *any* H-orbit divides $|H| = p^e$ and hence it is a power of p. Thus, in order to finish the proof of Part (2), we really have to examine whether we can have any other H-orbit $\mathcal{O}_i$, different from the trivial one $\{H\}$, consisting of a single element H_i or not.

For this, we consider the orbits of H_i's for H-action other than $H = H_1$. If there exists an orbit $\mathcal{O}_i$ of H_i under H-action on S by conjugation which consists of a single group H_i only, then since H_i is not moved by conjugation by any elements of H, H is contained the normalizer N_i of H_i in G. Conversely, if H is contained in N_i, then H_i is fixed under the H-action on S by conjugation, and hence the orbit $\mathcal{O}_i$ of H_i consists of H_i alone. Thus, we have proved that an orbit $\mathcal{O}_i$ of H_i consists of H_i alone if and only if H is contained in the normalizer N_i of H_i in G.

Thus, if the H-orbit $\mathcal{O}_i$ of an H_i other than H consists of H_i alone, then both H and H_i are contained in N_i; moreover they are Sylow p-subgroups of N_i because p^e divides $|N_i|$ and $|N_i|$ divides $|G|$. Thus by Part (2) of Corollary 4.16, $H = gH_ig^{-1}$ for some $g \in N_i$, from this we finally conclude that $H = H_i$, because H_i is normal in N_i. Thus, the set S has only one H-orbit of order 1, $\{H\}$, and this finishes the whole proof of Part (2). □

COROLLARY 4.17. *s divides $|G|$.* □

COROLLARY 4.18. *Let H be a Sylow p-subgroup of a group G. Then there are no other H-orbits of order 1 in the set S of all Sylow p-subgroups of G, except for the trivial one, $\{H\}$, for the action of H on S by conjugation.* □

Exercises for Chapter 4.4

EXERCISE 4.28. *Consider the quaternion group G in Example 2.15.*

(1) *Prove that G has a center Z of order 2 and G/Z is isomorphic to the Klein 4 group.*
(2) *Show that G is not isomorphic to D_4.*

EXERCISE 4.29. *Prove that a non-abelian group of order 8 is isomorphic to either the quaternion group or D_4.*

EXERCISE 4.30. *Determine the number of elements of order 5 which are contained in a group of order 20.*

EXERCISE 4.31. *Prove that no group of order p^2q, where p and q are prime, is simple.*

EXERCISE 4.32. *Let G be a group of order $p^e m$, where p prime and $(p, m) = 1$. Prove that G has a subgroup of order p^r, for each $r \leq e$.*

EXERCISE 4.33. *Let G be a group of order $p^e m$, where p prime and $(p, m) = 1$, and $1 \leq m < p$. Prove that G has a proper normal subgroup.*

5. Applications of Sylow Theorems

PROPOSITION 4.6. *(Classification of groups of order 6) There are exactly two isomorphism classes of groups of order 6. They are the classes of the cyclic groups C_6 and of the dihedral group $D_3 \simeq S_3$.*

PROOF. By the First Sylow Theorem (Theorem 4.6) a group G of order 6 contains an element x of order 3 and an element y of order 2. Then the 6 products $x^i y^j$ with $0 \leq i < 3, 0 \leq j < 2$ are all distinct element of G. This is true because if $x^i y^j = x^r y^s$, then $x^{i-r} = y^{s-j}$. Then since every power of x except the identity has order 3 and every power of y except the identity has order 2, $x^{i-r} = y^{s-j} = 1$, which shows that $i = r$ and $j = s$. Since the order of G is 6, these 6 products are all the elements of G, and hence the product yx must be one of them, and this easily leads us to have

$$yx = xy \qquad \text{or} \qquad yx = x^2y.$$

Then either of these two relations, together with $x^3 = 1$ and $y^2 = 1$, will enable us to have the multiplication table for the group G. Therefore, there are at most two isomorphism classes of groups of order 6. But, we already know the two isomorphism classes, namely, the classes of the cyclic group C_6 and of the dihedral group D_3. Thus, they are the only ones. □

PROPOSITION 4.7. *(Classification of groups of order 15) Every group G of order 15 is cyclic.*

PROOF. $15 = 3\cdot 5$, so by the Third Sylow Theorem (Theorem 4.8), the number of Sylow 3-subgroups of G divides 5 and is congruent $1 \pmod 3$; hence there is only one Sylow 3-subgroup H of G. Then by Corollary 4.10, it is a normal subgroup of G. Similarly, we see that there exist only one Sylow 5-subgroup K of G which is also normal in G. Then their intersection $H \cap K = \{1\}$, because its order divides both 5 and 3, and their product KH, which is a subgroup G by Part (2) of Proposition 2.16, has order > 5; hence $KH = G$. Thus, by Part (3) of Proposition 2.16, G is isomorphic to the product group $H \times K$. This group is cyclic of order 15 by Proposition 2.15; hence every group of order 15 is isomorphic to C_{15}. □

PROPOSITION 4.8. *(Classification of groups of order 21) There are two isomorphism classes of groups of order 21. They are:*

(1) *The class of the cyclic group C_{21};*
(2) *The class of the group G generated by two elements x, y satisfying the 3 relations*

$$x^7 = 1, y^3 = 1, yx = x^2y.$$

PROOF. $21 = 3\cdot 7$, so by the Third Sylow Theorem (Theorem 4.8), the number of Sylow 7-subgroups of G divides 3 and is congruent $1 \pmod 7$;

hence there is only one Sylow 7-subgroup K of G. Then by Corollary 4.10, it is a normal subgroup of G. But there can be more than one Sylow 3-subgroups of G. Let H be a 3-Sylow subgroup of G. Since both K and H are cyclic, let x be the generator of K and y the generator of H. Thus, $x^7 = 1$ and $y^3 = 1$. Moreover, since K is normal in G, $yxy^{-1} = x^i$ for some $i < 7$. Then since $y^3 = 1$, we have:

$$x = y^3xy^{-3} = y^2x^iy^{-2} = y(x^i)^iy^{-1} = ((x^i)^i)^i = x^{i^3}.$$

Thus, by taking down the powers, we see that i satisfies $i^3 \equiv 1 (\bmod 7)$. Since $i < 7$, $i = 1, 2$, and 4. This means that we have to argue three cases:

Case 1: $i = 1$. In this case, we have $yxy^{-1} = x$; hence G is abelian by Part (2) of Proposition 2.16. Since any subgroup of an abelian group is automatically normal, K and H are normal in G and $K \cap H = \{1\}$. Thus, by Part (3) of Proposition 2.16, it is isomorphic to a direct product group $H \times K$; hence it is a cyclic group of order 21 by Proposition 2.15.

Case 2: $i = 2$. In this case, we have $yxy^{-1} = x^2$. Then we have two generators x, y and three relations:

$$x^7 = 1, y^3 = 1, yx = x^2y.$$

Then one can easily verify that these relations enables us to rewrite every products of powers of x, y in the form

$$x^iy^j, \qquad 0 \leq i < 7, 0 \leq j < 3,$$

so that they are all the elements of G. We leave to the readers the details of constructing the multiplication table, consisting of x^iy^j, for G as an Exercise.

Case 3: $i = 4$. In this case, we have $yxy^{-1} = x^4$. We know that y^2 is also a generator of H, so we may replace y with y^2. Then

$$y^2xy^{-2} = yx^4y^{-1} = x^{16} = x^2,$$

thus this case is reduced to the previous one.

This finishes the whole proof. $\square$

PROPOSITION 4.9. *(Classification of groups of order 12) There are 5 isomorphism classes of groups of order 12. They are:*

(1) *The abelian group,* $C_3 \times C_4 \simeq C_{12}$;
(2) *The abelian group,* $C_2 \times C_2 \times C_3 \simeq C_2 \times C_6$;
(3) *The alternating group,* A_4;
(4) *The dihedral group,* D_6;
(5) *The group generated by two elements,* x, y *with relations*

$$x^4 = 1, y^3 = 1, xyx^{-1} = y^2.$$

PROOF. Since $12 = 2^2 \cdot 3$, G has a Sylow 2-subgroup H of order 4 and a Sylow 3-subgroup K of order 3. Thus, K is a cyclic group and H is either a cyclic group of order 4 or the Klein 4-group $V \simeq C_2 \times C_2$:

$$H \simeq C_4 \quad \text{or} \quad H \simeq C_2 \times C_2. \tag{4.13}$$

By the Third Sylow theorem (Theorem 4.8), we know the number of Sylow 2-subgroups is either 1 or 3 and the number of Sylow 3-subgroups is either 1 or 4. We then recall that a subgroup L of a group G is normal if and only if its obit under conjugation consists of L alone (Corollary 4.10). Since 12 is a fairly small number we have:

CLAIM 4.1. *Either* H *or* K *is a normal subgroup of* G.

PROOF OF CLAIM. Suppose that K is not normal in G. Then by Corollary 4.10, it has 4 conjugate subgroups $K = K_1, ..., K_4$. Since the order of K_i is 3, $K_i \cap K_j = \{1\}$ for any $1 \leq i \neq j \leq 4$. Thus,

$$G - \bigcup_{i=1}^{4} K_i$$

consists of 3 elements, say a, b, c. Then since *any* Sylow 2-subgroup H has order 4 and $H \cap K_i = \{1\}$ for each $i = 1, 2, 3, 4$, we must have:

$$H = \{1, a, b, c\}.$$

Thus, we have only one Sylow 2-subgroup H. Then again by Corollary 4.10, H is normal in G. □

Thus, by Part (2) of Proposition 2.16, we know that the product HK is a subgroup of G. Now, since $H \cap K = \{1\}$, we see that the 12 elements hk are all distinct; hence $HK = G$.

Case 1: Both H and K are normal in G. In this case, G is isomorphic to $H \times K$ by Part (3) of Proposition 2.16. Thus, by (4.13), we see that there are two isomorphism classes:

$$G \simeq C_4 \times C_3 \qquad \text{or} \qquad G \simeq C_2 \times C_2 \times C_3.$$

This establishes Part (1) and Part (2).

Case 2: H is a normal subgroup of G, but K is not. In this case, by Corollary 4.10, we have 4 conjugate Sylow 3-subgroups, $S = \{K = K_1, ..., K_4\}$. Let G operate on S by conjugation. Then by Proposition 3.6, this action determines a permutation presentation

$$\varphi : G \longrightarrow S_4,$$

by sending

$$g \longmapsto m_g,$$

where $m_g(s) = g.s = gsg^{-1}$ for any $s \in S$. In what follows, we are going to prove that φ is injective and its image is A_4, thus proving that G is isomorphic to A_4:

We recall that the stabilizer of K_i for the action of conjugation is, by definition, the normalizer N_i of K_i, and K_i is automatically contained in N_i. Since K is not a normal subgroup of G, N_i is a proper subgroup of G by Corollary 4.11, and hence $|N_i| = 3$, because K_i is a subgroup of N_i by the counting formula 4.13. Thus, $N_i = K_i$ for each $i = 1, 2, 3, 4$. On the other hand, since the identity element 1 is the only common element of these 4 distinct subgroups K_i, left multiplication m_g by g, sending

$$K_i \longmapsto g.K_i, = gK_ig^{-1},$$

stabilizes K_i only when $g = 1$. This is because the 4 subgroups K_i are different. Thus, m_g is faithful, i.e., φ is injective.

Now, we are left to prove that the image of φ is A_4. Since we have 4 different cyclic subgroups K_i of order 3, we have $(3-1) \times 4 = 8$ elements of order 3; and these 8 elements certainly generates the group G, because the order of the subgroup of G generated by these 8 elements is ≥ 8 and divides $12 = |G|$ by Lagrange's Theorem (Corollary 2.8). Now, let $x \in G$ be an element of order 3. Then its image $\varphi(x)$ in S_4 has the same order 3, because φ is injective. Thus $\varphi(x)$ is an even permutation, because any permutation of order 3 is even: $(123) = (12)(13)$. Thus, all the images of these 8 elements are in A_4. Therefore, $G \simeq A_4$. This establishes Part (3).

Case 3: K is a normal subgroup of G, but H is not. In this case, let H act on K by conjugation. Then since K is normal in G, conjugation hkh^{-1} by an element h of H is in K; and hence it is automatically an automorphism $k \longmapsto hkh^{-1}$ of K. Let y be the generator for the cyclic subgroup K. Then

$$K = \{1, y, y^2\},$$

with $y^3 = 1$. We then note that there are only two automorphisms of K, the identity map and the one which switches y and y^2.

On the other hand, we have two possibilities for H: either H is cyclic of order 4 or is isomorphic to the Klein 4 group.

Suppose that H is cyclic of order 4. Let x be its generator. Thus, $x^4 = 0$. Then since H is not normal in G, $G = HK$ is not an abelian group; hence $xy \neq yx$, i.e., conjugation by x is not the trivial automorphism of K. Thus,

$$xyx^{-1} = y^2.$$

Then one can verify that x, y with the three relations:

$$x^4 = 1, y^3 = 1, xyx^{-1} = y^2,$$

do generate a subgroup of order 12 of G; thus, it must be G. This establishes Part (5).

Now finally, suppose that H is isomorphic to the Klein 4 group. Then since there are only 2 automorphisms of K, there exists an element $w \in H$ of order 2 which acts trivially:

$$wyw^{-1} = y.$$

On the other hand, since G is not abelian, there exists $v \in H$ of order 2 which acts non-trivially:

$$vyv^{-1} = y^2.$$

Thus, $H = \{1, v, w, vw\}$ and the relations

$$v^2 = 1, w^2 = 1, vw = wv$$

hold in H. Then one easily verify that the element

$$x := wy \in G$$

has order 6. Moreover, since $x^{-1} = y^2w$, we have

$$vxv^{-1} = vwyv^{-1} = wvyv^{-1} = wy^2 = ywy = y^2w = x^{-1}.$$

Thus, $x, v \in G$ with the three relations

$$x^6 = 1, v^2 = 1, vx = x^{-1}v$$

define a subgroup D_6 of order 12 of G; hence $G = D_6$. This establishes Part (4). □

Exercises for Chapter 4.5

EXERCISE 4.34. *Determine the Class Equation for the groups of order* 12 *in Proposition 4.9.*

EXERCISE 4.35. *Find all groups of order* 20 *up to isomorphism.*

EXERCISE 4.36. *Prove that any group of order* 10 *is isomorphic to* Z_{10} *or* D_5.

EXERCISE 4.37. *Determine all groups of order* $2p$, *where* p *is a prime number.*

EXERCISE 4.38. *Prove that the cyclic group of order* p *is the only proper normal subgroup of* D_p, *where* p *is an odd prime number.*

EXERCISE 4.39. *Prove that the only simple groups of order* < 60 *are groups of prime order.*

EXERCISE 4.40. *Find all groups of order* 33 *up to isomorphism.*

EXERCISE 4.41. *Find all groups of order* 18 *up to isomorphism.*

EXERCISE 4.42. *Prove those claims in Example 2.36.*

CHAPTER 5

Rings

1. The Definition of a Ring

1.1. Motivations. In Chapter One, we introduced a few basic definitions in Algebra, including the definition of a ring. In studying rings, the set $\mathbb{Z}$ of all integers is our basic model for the concept of a ring. They are closed under addition, subtraction, and multiplication, but not division. We recall that any field is a commutative ring, so that the set $\mathbb{C}$ of all complex numbers is a ring. An easy way of having a few examples of rings is to consider subrings of $\mathbb{C}$.

DEFINITION 5.1. *A* ***subring*** *of $\mathbb{C}$ is a subset which is closed under addition, subtraction, and multiplication and moreover which contains the multiplicative identity 1 of $\mathbb{C}$.*

REMARK 5.1. *Such a subring of $\mathbb{C}$ does exist, because $\mathbb{C}$ itself is such a subring in the worst case.*

EXAMPLE 5.1. *Any subfield is a subring, so it is a ring.*

EXAMPLE 5.2. *The ring $\mathbb{Z}[i]$ of Gauss integers,*

$$\mathbb{Z}[i] = \{a + bi : a, b \in \mathbb{Z}\},$$

where $i = \sqrt{-1}$, is a subring of $\mathbb{C}$, so it is ring, because it is closed under addition, subtraction, multiplication, and it contains 1. We note that $\mathbb{Z}[i]$ represents a lattice in the complex plane.

In fact, we can form a subring $\mathbb{Z}[\alpha]$ in an analogous manner to the ring of Gauss integers:

DEFINITION 5.2. *Let* α *be a complex number and define*

$$\mathbb{Z}[\alpha]$$

to be the smallest *subring of* $\mathbb{C}$ *containing both* $\mathbb{Z}$ *and* α*, and we call such a ring the* ***subring*** *of* $\mathbb{C}$ ***generated by*** α ***over*** $\mathbb{Z}$*.*

Since $\mathbb{Z}[\alpha]$ is by definition a ring containing α, it will contain all the powers α^j of α because it is closed under multiplication. Also, it will contain sums and differences of such powers, and it obviously contains 1. Thus, it contains every complex number β which can be expressed as a polynomial in α with integer coefficients:

$$\beta = a_n\alpha^n + \cdots + a_1\alpha + a_0, \tag{5.1}$$

where each $a_i \in \mathbb{Z}$. Moreover, the set of all such numbers is closed under addition, subtraction, and multiplication, and it contains 1. Thus, it is the subring of $\mathbb{C}$ generated by α over $\mathbb{Z}$.

EXAMPLE 5.3. *The ring* $\mathbb{Z}[\frac{1}{2}]$ *consists of those rational numbers which can be expressed as a polynomial in* $\frac{1}{2}$ *with integer coefficients. Then since these rational numbers are simply those whose denominator is a power of* 2, $\mathbb{Z}[\frac{1}{2}]$ *is not represented as a lattice in the complex plane.* $\mathbb{Z}[\frac{1}{2}]$ *is in fact a dense subset of the real line* $\mathbb{R}$*.*

There are two kinds of complex numbers:

DEFINITION 5.3. *Let* α *be a complex number.*

1. *The number* α *is* ***algebraic*** *(over* $\mathbb{Z}$*) if it is a root of a polynomial with integer coefficients.*
2. *The number* α *is* ***transcendental*** *(over* $\mathbb{Z}$*) if it is never a root of any polynomial with integer coefficients.*

EXAMPLE 5.4. $i = \sqrt{-1}$, $i+3$, $1/2$, $2+\sqrt[3]{7}$, $\sqrt{5}+\sqrt{-5}$ *are algebraic numbers, whereas the numbers* e *and* π *are transcendental numbers, although it is not easy to prove that they are.*

REMARK 5.2. *We note that a transcendental number is always irrational, since any rational number is algebraic. History tells us the followings: J.H. Lambert (1728 - 1777) was the first to prove in 1768 that π is irrational. Lambert also conjectured in the same paper, where he proved that π is irrational, that e and π are both transcendental.*

It then took more than a hundred years for their first proofs: C. Hermite (1822 - 1901) was the first to prove in 1873 that the number e is transcendental. Using methods similar to those of Hermite, Lindemann (1852 - 1939) was the first to prove in 1882 that the number π is transcendental.

Let α be a complex number. Then we have two cases, that is, either transcendental or algebraic but not both, according to the above definition:

Case 1. If a complex number α is transcendental, then any two distinct polynomial expressions (5.1) should represent different complex numbers, otherwise it is algebraic; hence we have a one-to-one correspondence between the ring $\mathbb{Z}[\alpha]$ and the set $\mathbb{Z}[x]$ of integer polynomials $p(x)$ in the unknown x, by the map sending $p(\alpha) \longleftrightarrow p(x)$. In a later section, we are going to formally construct the polynomial ring $\mathbb{Z}[x]$ over $\mathbb{Z}$.

Case 2. On the other hand, if a complex number α is algebraic, then there could be many polynomial expressions (5.1) which represent the same complex number.

EXAMPLE 5.5. *We consider $\alpha = i$ $(=\sqrt{-1})$. Then the powers α^n are either ± 1 or $\pm i$.*

In fact, having many polynomial expressions for an algebraic number is very good, since it allows us to simplify its polynomial expression (5.1) to the one with the least degree in α.

EXAMPLE 5.6. *Thus with the relation $i^2 = -1$, we can reduce any polynomial expressions (5.1) to the one whose degree in i is less than or equal to 1, and this completely agrees with the definition of the ring $\mathbb{Z}[i]$ of the Gauss integers.*

1.2. The Definition of a Ring. In Chapter One, we have already defined a ring. The definition of an abstract ring is similar to that of a field, except that multiplicative inverses are not required to exist:

DEFINITION 5.4. *A* ***ring*** *R is a set with two laws of composition, denoted by $+$ and $\cdot$, called addition and multiplication, satisfying the following axioms:*

(1) *With the law of composition $+$, R is an abelian group, with identity element denoted by 0.*
(2) *Multiplication is associative and has an identity denoted by 1.*
(3) *(Distributive laws) For each $a, b, c \in R$, the following two equalities hold:*

$$(a+b)c = ac + bc \qquad \text{and} \qquad a(b+c) = ab + ac.$$

We have an immediate

LEMMA 5.1. *Let R be a ring. Then for any $a \in R$,*

$$0a = 0 \qquad \text{and} \qquad a0 = 0.$$

PROOF. We note that

$$0a + 0a = (0+0)a = 0a = 0a + 0,$$

and hence by cancelling out $0a$ from both sides, we conclude that $0a = 0$. Similarly, we see that $a0 = 0$. □

LEMMA 5.2. *Let R be a ring. Then for any $a \in R$,*

$$-(-a) = a.$$

PROOF. Since $-a+a = 0$ and $-a+(-(-a)) = 0$, we see that $-(-a) = a$ by uniqueness of the additive inverse of $-a$ in R. □

LEMMA 5.3. *Let a, b be elements of a ring R. Then*

$$(-a)b = -(ab) = a(-b).$$

PROOF. We note that

$$\begin{aligned} 0 &= 0b = (a + (-a))b = ab + (-a)b, \\ 0 &= ab + (-(ab)). \end{aligned}$$

Thus, we see that $(-a)b = -(ab)$, by uniqueness of the additive inverse of ab. Similarly, $a(-b) = -(ab)$ holds. □

COROLLARY 5.1. $(-a)(-b) = ab$.

PROOF. $(-a)(-b) = -a(-b) = -(-(ab)) = ab$. □

COROLLARY 5.2. $-a = (-1)a$.

PROOF. $a + (-1)a = 1a + (-1)a = (1 + (-1))a = 0a = 0$, hence $(-1)a = -a$ by uniqueness of the additive inverse of a. □

LEMMA 5.4. *(General Distributive Law) For elements $a_1, ..., a_m$ and $b_1, ..., b_n$ of a ring R, we have*

$$(a_1 + \cdots + a_m)(b_1 + \cdots + b_n) = \sum_{i=1}^{m} \sum_{j=1}^{n} a_i b_j.$$

PROOF. For $m = 1$, we prove the equality by induction on n: if $n = 2$, then $a_1(b_1 + b_2) = a_1b_1 + a_1b_2$ by one of the distributive law. Then

$$\begin{aligned} a_1(b_1 + \cdots + b_n + b_{n+1}) &= a_1(b_1 + \cdots + b_n) + a_1b_{n+1} \\ &= a_1b_1 + \cdots + a_1b_n + a_1b_{n+1}. \end{aligned}$$

Thus, the equality is true for $m = 1$ and for all $n \in \mathbb{N}$. With this being proved for $m = 1$ and for all n, we then apply a second induction on m: If $m = 2$, then

$$\begin{aligned} (a_1 + a_2)(b_1 + \cdots + b_n) &= a_1(b_1 + \cdots + b_n) + a_2(b_1 + \cdots + b_n) \\ &= a_1b_1 + \cdots + a_1b_n + a_2b_1 + \cdots + a_2b_n. \end{aligned}$$

Now,

$$\begin{aligned}(a_1 + \cdots + a_m + a_{m+1})(b_1 + \cdots + b_n) &= (a_1 + \cdots + a_m)(b_1 + \cdots + b_n) \\ &\quad + a_{m+1}(b_1 + \cdots + b_n) \\ &= \cdots \\ &= \sum_{i=1}^{m+1} \sum_{j=1}^{n} a_i b_j.\end{aligned}$$

Thus, the equality holds for all m and all n. □

REMARK 5.3. *For a positive integer n, we recall*

$$na := \underbrace{a + \cdots + a}_{n \text{ times}}$$

for any $a \in R$ (Notation 2.6).

LEMMA 5.5. *For all integers n and all $a, b \in R$, we have*

$$n(ab) = (na)b = a(nb).$$

PROOF. For a positive integer n,

$$\begin{aligned} n(ab) &:= \underbrace{ab + \cdots + ab}_{n \text{ times}} \\ &= (a + \cdots + a)b \\ &= (na)b, \end{aligned}$$

by the general distributive law. Similarly, $a(nb) = n(ab)$ holds.

Now, for a negative integer n, we apply Lemma 5.3. □

DEFINITION 5.5. *A* ***subring*** *of a ring R is a subset of R which is closed under the operations of addition, subtraction, and multiplication, and it contains the identity element 1.*

REMARK 5.4. *Some old people, mainly working on the Theory of Rings as an isolated subject of study, do not usually require the existence of a multiplicative identity in a ring. But, we do require its existence, at least because of the following two reasons. We require our ring to have the underlying structure of a monoid, and moreover, we want to have a functorial correspondence, for instance, the one in Algebraic Geometry, that is, a contravariant functor from the category of commutative rings R with identity* 1 *and the category of geometric objects, Spec*(R).

DEFINITION 5.6. *A* ***commutative ring*** *is a ring whose multiplication is commutative.*

EXAMPLE 5.7. *Not all rings are commutative. For instance, the set* $\mathbb{R}^{n\times n}$ *of* $n\times n$ *real matrices is a prototypical example of a non-commutative ring with multiplicative identity* I*, the identity matrix, under matrix addition and matrix multiplication.*

NOTATION 5.1. *From now on, our rings are going to be commutative, unless we otherwise mention explicitly non-commutativity of a ring.*

Besides subrings of $\mathbb{C}$, the most important rings are polynomial rings. Given any ring R, a polynomial in an unknown x with coefficients in R is an expression of the form

$$a_n x^n + \cdots + a_1 x + a_0,$$

where each $a_i \in R$. The set of these polynomials forms a ring, denoted by

$$R[x].$$

We are going to construct the polynomial ring in the next section.

EXAMPLE 5.8. *Here are some examples of rings other than subrings of* $\mathbb{C}$.

(1) *Any field is a ring.*
(2) *The set of all continuous real-valued functions of a real variable* x *form a ring, with addition and multiplication of functions:*

$$(f+g)(x) = f(x) + g(x) \qquad \text{and} \qquad (fg)(x) = f(x)g(x).$$

(3) *The zero ring* $R = \{0\}$ *consists of a single element* 0.
(4) *The quotient group* $\mathbb{Z}/n\mathbb{Z}$ *becomes a ring under the induced operations* $+$ *and* $\cdot$ *from those on* $\mathbb{Z}$*, and this type of ring is called a* ***quotient ring****.*

In particular, if n is not a prime number then the quotient ring $\mathbb{Z}/n\mathbb{Z}$ has "zero-divisors". For instance, if $n = 6 = 2 \cdot 3$, $\bar{2} = 2$ and $\bar{3} = 3$ are zero-divisors in the ring $\mathbb{Z}/6\mathbb{Z}$, because

$$\bar{2} \cdot \bar{3} = \overline{2 \cdot 3} = \bar{6} = \bar{0} = 0,$$

holds in $\mathbb{Z}/6\mathbb{Z}$.

REMARK 5.5. *In a field F, every non-zero element is invertible: hence $F - \{0\}$ is a group under multiplication; hence any field has at least two distinct elements, 0 and 1, hence these two cannot be the same in a field. But in a ring we could have a relation, $0 = 1$.*

LEMMA 5.6. *Let R be a ring in which $0 = 1$. Then R is the zero ring $\{0\}$.*

PROOF. If $1 = 0$ in R, then for any $a \in R$,

$$a = 1a = 0a = 0.$$

$\square$

Although multiplicative inverses are not required to exist in a ring R, some particular elements of R may have an inverse, and the inverse is unique if it exists. Elements which have multiplicative inverse are called **units**.

EXAMPLE 5.9. *The units in $\mathbb{Z}$ are -1 and $+1$, and the units in the polynomial ring $\mathbb{R}[x]$ in one unknown x over $\mathbb{R}$ are the non-zero constant polynomials, i.e., the non-zero real numbers. In a field, every non-zero element is a unit.*

The identity element 1 of a ring R is always a unit, and any reference to "the" unit element of R refers to the identity element 1, and it is too late to change such a reference.

Exercises for Chapter 5.1

EXERCISE 5.1. *Describe the group of units in each ring:*

(1) $\mathbb{Z}/5\mathbb{Z}$,
(2) $\mathbb{Z}/6\mathbb{Z}$.

EXERCISE 5.2. *Show that* $5+\sqrt[3]{2}$ *is an algebraic number.*

EXERCISE 5.3. *Prove that* $\cos(2\pi/n)$ *is an algebraic number for each integer* n. *For example, when* $n=5$, *verify that*

$$\cos(2\pi/5) = \mathrm{Re}\left(\frac{\sqrt{5}-1}{4} + \frac{i}{4}\sqrt{2(5+\sqrt{5})}\right)$$

EXERCISE 5.4. *Prove that* $\mathbb{Q}[\sqrt{2},\sqrt{3}] = \mathbb{Q}[\sqrt{2}+\sqrt{3}]$, *where the first ring is the smallest subring of* $\mathbb{C}$ *containing* $\sqrt{2},\sqrt{3}$, *and* $\mathbb{Q}$. *[Hint: By squaring* $a := \sqrt{2}+\sqrt{3}$, *we see that* $\sqrt{6}\in\mathbb{Q}[\sqrt{2}+\sqrt{3}]$. *Then by multiplying it to* $\sqrt{2}+\sqrt{3}$, *we see that* $b := 2\sqrt{3}+3\sqrt{2}\in\mathbb{Q}[\sqrt{2}+\sqrt{3}]$. *Thus,* $\sqrt{2} = b-2a \in \mathbb{Q}[\sqrt{2}+\sqrt{3}]$. *Similarly,* $\sqrt{3}\in\mathbb{Q}[\sqrt{2}+\sqrt{3}]$*].*

EXERCISE 5.5. *Prove that the units, i.e., the invertible elements, of* $\mathbb{Z}[i]$ *are* $\pm 1, \pm i$.

EXERCISE 5.6. *Prove that* $1+x$ *is a unit in* R *if* x *in a ring* R *satisfies that* $x^n = 1$ *for some* $n > 1$.

2. Abstract Construction of Integers and Polynomials

In this section, we will sketch how we can construct the set $\mathbb{N}$ of natural numbers and try to prove the associative and distributive laws for numbers, based on Peano's axioms for it. This is due to G. Peano (1858 - 1932):

DEFINITION 5.7. *The set* $\mathbb{N}$ *of all natural numbers* n *is* characterized *by the following conditions, called* ***Peano's axioms****:*

(1) *The set* $\mathbb{N}$ *has a particular element* 1.
(2) *(Successor function). There exists an* injective *map* $\sigma : \mathbb{N} \longrightarrow \mathbb{N}$ *that sends each natural number* $n\in\mathbb{N}$ *to another number, called the next number or* ***successor*** *and* $\sigma(n)\neq 1$ *for any* $n\in\mathbb{N}$.
(3) *(Induction Axiom). Suppose that a subset* S *of* $\mathbb{N}$ *satisfies the following properties:*

(a) $1 \in S$;

(b) *if* $n \in S$ *then* $\sigma(n) \in S$.

Then S *contains every natural number, namely,* $S = \mathbb{N}$.

REMARK 5.6. *Since addition of two natural numbers is not yet defined, we do not write* $\sigma(n) = n+1$. *Rather than this, it is better to denote* $\sigma(n)$ *by* n' *now. But, once addition is defined we will denote* $\sigma(n)$ *by* $n+1$.

REMARK 5.7. *The successor function allows us to use the natural numbers for counting, which is the basis of arithmetic.*

Property (3) in the above definition is the *induction property* of the natural numbers, and it is called "the **Principle of Mathematical Induction**". Intuitively, it says that the natural numbers are obtained from 1 by repeatedly taking the next integer:

$$\mathbb{N} = \{1, 1', 1'', ...\} = \{1, 2, 3, ...\},$$

that is, counting runs through all natural numbers, and this property is the formal basis of induction proofs.

Suppose that we want to prove that a statement $P(n)$ is true for every natural number $n \in \mathbb{N}$. For this, we let S be the set of natural numbers for which $P(n)$ is true. To say that $P(n)$ is true for every natural number is exactly the same as saying that $S = \mathbb{N}$. Then the third condition, Induction Axiom or Induction principle, in the above Definition is equivalently translated into the following statement:

$$\left.\begin{array}{c} P(1) \text{ is true,} \\ \text{If } P(n) \text{ is true, then } P(n') \text{ is true.} \end{array}\right\} \Longrightarrow \text{ Then } S = \mathbb{N}.$$

EXAMPLE 5.10. *When Gauss (1777 - 1855) was a student in an Elementary School, he figured out that the sum of n numbers,* $1, 2, ..., n$ *is equal to* $n(n+1)/2$ *by adding the same sum twice in a different order and dividing the result by* 2*:*

$$\begin{array}{rcl} S_1(n) & = & 1 + \quad 2 \quad + \cdots + n, \\ S_1(n) & = & n + (n-1) + \cdots + 1, \end{array}$$

so that $2 \cdot S_1(n) = n(n+1)$*. Then applying the Induction Principle, one can prove that*

$$S_1(n) = n(n+1)/2$$

for each $n \in \mathbb{N}$*.*

REMARK 5.8. *We should note that the Induction Principle itself never tells us how one originally gets the formula, like*

$$1 + 2 + \cdots + n = n(n+1)/2.$$

For example, can you derive the formula for

$$S_2(n) = 1^2 + 2^2 + \cdots + n^2?$$

You will quickly recognize the Gauss's method does not work in this case, because the terms in this sum are not increasing in a linear manner which enables us to take the average of all the terms.

Once the formula for $S_2(n)$ *is derived, then we just apply the Induction Principle to conclude that the formula for* $S_2(n)$ *holds for every* $n \in \mathbb{N}$*.*

2.1. Addition and Multiplication of Natural Numbers.

DEFINITION 5.8. *For a natural number* n*, we denote* $\sigma(n)$ *by*

$$n + 1.$$

EXAMPLE 5.11. $\sigma(1) = 1 + 1\ (= 2)$, $\sigma(2) = 2 + 1\ (= 3)$, *etc.*

Then since the successor function σ is injective by definition, we know that $\sigma^{-1}(2) = 1$, $\sigma^{-1}(3) = 2$, so that $\sigma^{-1}(\sigma^{-1}(3)) = 1$; hence for any natural number $n \neq 1$,

$$\sigma^{-1}(\sigma^{-1}(\cdots(\sigma^{-1}(n)))) = 1,$$

where the inverse map

$$\sigma^{-1} : \mathbb{N} - \{1\} \longrightarrow \mathbb{N}$$

is applied $\sigma^{-1}(n)\ (= n - 1)$ times. That is, for any natural number n, we have

$$n = \sigma^{-1}(\sigma^n(1)),$$

where σ^n is the composition of σ n times: $\sigma^n = \underbrace{\sigma \circ \cdots \circ \sigma}_{n \text{ times}}$. Thus, $\sigma(n) = \sigma^n(1)$.

This allows us to define **addition of two natural numbers**:

DEFINITION 5.9. *For natural numbers m and n, define*

$$m + n := \sigma^{-1}(\sigma(\cdots(\sigma(\sigma(1))))) = \sigma^{-1}(\sigma^m(\sigma^n(1))).$$

Now, if $m = 1$ and n is an arbitrary natural number, then

$$1 + n = \sigma^n(1) = n + 1,$$

where we have the first equality by Definition 5.9 and the last equality by definition of $\sigma(n)$. Moreover, for any m and n, we have

$$\begin{aligned} m + n &= \sigma^{-1}(\sigma^m(\sigma^n(1))) = \sigma^{-1}(\sigma(\sigma(\cdots(\sigma(1))))) \\ &= \sigma^{-1}(\sigma^n(\sigma^m(1))) = n + m, \end{aligned}$$

and this proves that the **binary operation** $+$ defined in Definition 5.9 is *commutative*: For any natural numbers m and n, we have

$$m + n = n + m.$$

Moreover, we have

$$\begin{aligned} k + (m + n) &= \sigma^{-1}(\sigma^k(\sigma^{m+n}(1))) = \sigma^{-1}(\sigma(\sigma(\cdots(\sigma(1))))) \\ &= \sigma^{-1}(\sigma^{k+m}(\sigma^n(1))) = (k + m) + n, \end{aligned}$$

and this proves that *addition is associative*. Thus,

$$\begin{aligned} 2 + n &= (1 + 1) + n = 1 + (1 + n) \\ &= 1 + (n + 1) = (1 + n) + 1 \\ &= (n + 1) + 1 = n + (1 + 1) = n + 2. \end{aligned}$$

Now suppose that

$$k + m = k + n.$$

If $k = 1$, then $\sigma(m) = 1 + m = 1 + n = \sigma(n)$; hence $m = n$ since σ is injective. If $k \neq 1$, then

$$\sigma^{-1}(\sigma^k \sigma^m(1)) = \sigma^{-1}(\sigma^k \sigma^n(1)),$$

and hence

$$\sigma^k(\sigma^{m-1}(1)) = \sigma^k(\sigma^{n-1}(1)).$$

Thus,

$$m = \sigma^{m-1}(1) = \sigma^{n-1}(1) = n,$$

because σ^k is injective because σ is injective, and this proves that *the cancellation law holds for addition*:

$$k + m = k + n \qquad \Longrightarrow \qquad m = n$$

for all natural numbers k, m, and n.

Moreover, by Definition 5.9 we have

$$\begin{aligned} n + n &= \sigma^n(n) = \sigma^{-1}(\sigma^n(\sigma^n(1))) \\ n + (n + n) &= \sigma^n(\sigma^n(n)) = \sigma^{-1}(\sigma^n(\sigma^{(n+n)}(1))) = \sigma^{-1}(\sigma^n(\sigma^n(\sigma^n(1)))), \\ & \cdots \end{aligned}$$

and hence this allows us to define **multiplication of two natural numbers**:

DEFINITION 5.10. *For any natural number m and n, define*

$$mn := \sigma^{-1}(\sigma^n(\sigma^n \cdots (\sigma^n(1)))) = \sigma^{-1}((\sigma^n)^m(1)).$$

Now, if $m = 1$ and n is an arbitrary natural number, then

$$\begin{aligned} 1n &= \sigma^{-1}((\sigma^n)^1(1)) = \sigma^{-1}(\sigma^1)^n(1)) \\ &= n1, \end{aligned}$$

and this proves that 1 is the identity for multiplication $\cdot$, because $\sigma^1 = \sigma$ so that

$$1n = n1 = \sigma^{-1}((\sigma^1)^n(1)) = \sigma^{-1}(\sigma^n(1)) = n.$$

Moreover, we have

$$\begin{aligned} mn &= \sigma^{-1}((\sigma^n)^m(1)) = \sigma^{-1}(\sigma(\cdots(\sigma(\sigma(1))))) \\ &= \sigma^{-1}((\sigma^m)^n(1)) = nm, \end{aligned}$$

and this proves that the **binary operation** $\cdot$ defined in Definition 5.10 is *commutative*: For any natural numbers m and n, we have

$$mn = nm.$$

Moreover, we have

$$\begin{aligned} k(mn) &= \sigma^{-1}((\sigma^{mn})^k(1)) = \sigma^{-1}(\sigma^{mnk}(1)) \\ &= \sigma^{-1}(\sigma(\cdots(\sigma(\sigma(1))))) = \sigma^{-1}((\sigma^{km})^n(1)) \\ &= (km)n, \end{aligned}$$

and this proves that *multiplication is associative.*

Finally, we have

$$\begin{aligned} k(m+n) &= \sigma^{-1}((\sigma^{(m+n)})^k(1)) = \sigma^{-1}(\sigma^{(m+n)k}(1)) \\ &= \sigma^{-1}(\sigma(\cdots(\sigma(\sigma(1))))) = \sigma^{-1}(\sigma^{km}(\sigma^{kn}(1))) \\ &= \sigma^{-1}(\sigma^{km+kn}(1)) = km + kn, \end{aligned}$$

and this proves that *multiplication is distributive over addition.*

2.2. Inequalities.

DEFINITION 5.11. *Let m and n be any natural numbers. we define*

$$m < n \qquad \text{if } n = m + k \text{ for some natural number } k.$$

We note that if such a k exists, then it is unique by the cancellation law for addition.

PROPOSITION 5.1. *(The Trichotomy Law) For any two natural numbers m and n, we have:*

$$m < n, \ m = n, \ \text{or } m > n,$$

but none of the two hold simultaneously.

PROOF. If $m = n$ then applying σ^{-1} successively several times to both sides, we get $1 = 1$; but if $m < n$, then $n = m+a$ for some natural number a, so applying σ^{-1} successively several time to both sides of $m < n$ we get $1 < 1+a = \sigma(a) \neq 1$, because 1 is never an image of any natural number under σ. Thus, it is impossible to have that both $m < n$ and $m = n$ hold.

Now if both $m < n$ and $n < m$ hold, then $n = m + a$ for some natural number a and $m = n + b$ for some natural number b; hence we have

$$n = n + (a + b).$$

Thus again applying σ^{-1} successively several times, we get

$$1 = 1 + (a + b) = \sigma(a + b),$$

but this is an absurd, because 1 is never an image of any natural number under σ, by definition. Thus, that both $m < n$ and $n < m$ hold never happens.

To finish the whole proof, it remains to prove by induction on n that for each m at least one of these three alternatives holds. If $n = 1$, then $m = 1$ or $m > 1$, thus it is O.K. when $n = 1$. Suppose that $n \neq 1$. By induction hypothesis we assume that the alternatives hold for any natural number m and for some one n. We need to show that for any natural number k and for $\sigma(n)$ the following holds:

$$k < \sigma(n), \qquad k = \sigma(n), \qquad \sigma(n) < k.$$

If $k = 1$, then the first holds. Otherwise, $k \neq 1$, so $k = \sigma(m)$ for some m; hence by induction hypothesis we have one of

$$m < n, \; m = n, \text{ or } \; n < m,$$

where $\sigma(m) = \sigma(n)$ implies $m = n$ because σ is injective. Now we then note, for instance, that $m < n$ implies $n = m+a$ for some natural number a; hence

$$\sigma(n) = \sigma(m + a) = \sigma(m) + a$$

holds. That is, $m < n$ implies $\sigma(m) < \sigma(n)$. Thus, we just apply σ to both sides of these inequalities,

$$m < n, \ m = n, \ \text{or} \ \ n < m,$$

and this finishes the whole proof. □

COROLLARY 5.3. *$m < n$ implies $\sigma(m) < \sigma(n)$.* □

PROPOSITION 5.2. *For any natural numbers k, m and n, we have*

$$\begin{aligned} m \ < \ n \quad &\Longrightarrow \quad m + k < n + k. \\ m \ < \ n \quad &\Longrightarrow \quad km < kn. \end{aligned}$$

PROOF. If $m < n$ then there exists a natural number a such that $n = m + a$. Thus, applying the same number of σ's to both sides of $n = m + a$, we have

$$\begin{aligned} n + k &= (m + a) + k \\ &= (m + k) + a, \end{aligned}$$

and hence $m + k < n + k$.

Now, since $n = m + a$ we have

$$\begin{aligned} kn &= \sigma^{-1}((\sigma^n)^k(1)) = \sigma^{-1}((\sigma^{(m+a)})^k(1)) \\ &= \sigma^{-1}(\sigma^{km+ka}(1)) = km + ka, \end{aligned}$$

and hence $km < kn$. □

COROLLARY 5.4. *A natural number k can be cancelled; namely,*

$$km = kn \qquad \Longrightarrow m = n.$$

PROOF. Otherwise, we have either $m > n$ or $m < n$. Thus by proposition, $km > kn$ or $km < kn$, any of which contradicts the hypothesis, $km = kn$. □

PROPOSITION 5.3. *For any natural numbers m and n, $n > m$ implies either $n = \sigma(m)$ or $n > \sigma(m)$ but not both.*

PROOF. If $n > m$ then there exists a natural number a such that $n = m + a$. If $a = 1$, then $n = m + 1 = \sigma(m)$. Otherwise, $a = \sigma(b)$ for some natural number b; hence $n = m + \sigma(b) = \sigma(m + b)$. But, $m + b > m$ always holds; hence $n = \sigma(m + b) > \sigma(m)$. □

COROLLARY 5.5. *There is no natural number between* 1 *and* $\sigma(1)$.

PROOF. If there exists a natural number n such that $1 < n < \sigma(1)$, then by the proposition, we have either $n = \sigma(1)$ or $n > \sigma(1)$; and any of which contradicts the hypothesis that $1 < n < \sigma(1)$. □

DEFINITION 5.12. *For any natural numbers m and n, we define*

$$m \leq n \qquad \text{if } m < n \text{ or } m = n.$$

PROPOSITION 5.4. *($\mathbb{N}$ is "well-ordered"). Every non-empty subset S of $\mathbb{N}$ contains a least element, that is, given a non-empty subset S of $\mathbb{N}$, there exists a natural number a such that $a \leq s$ for any $s \in S$, called a* ***least*** *element of S.*

PROOF. (By contradiction). Suppose that $\mathbb{N}$ has a non-empty subset S with no least element. Let $s \in S$ be *any* element of S. If we can prove that $s \geq n$ for each $n \in \mathbb{N}$ by induction on n, then a particular $n = s + 1$ leads us to a contradiction that $s \geq s + 1$ for each $s \in S$.

First, we see that when $n = 1$, $s \geq 1$ obviously holds. Now we suppose, as induction hypothesis, that $s \geq n$ for some n. Then $n \notin S$, for, if it were then we would have a least element n in S, which is not the case by hypothesis. Thus, in particular, $s \neq n$; hence $s > n$. Thus, by the above proposition, we conclude that $s \geq \sigma(n)$, which is what we wanted to prove to finish the proof. That is, we finish our proof by mathematical induction on n. □

DEFINITION 5.13. *Let S be a set.*

(1) *An order $<$ on S is called a* ***linear order*** *(or a* ***total order****) if the Trichotomy Law holds for $<$.*
(2) *An (totally)* ***ordered set*** *is a set with a linear order.*

(3) *An ordered set S is* ***well-ordered*** *if every non-empty subset of S contains a least element.*

EXAMPLE 5.12. *The above proposition states that $\mathbb{N}$ is a well-ordered set.*

AXIOM 1. *(Well-Ordering Principle) Every set S can be well-ordered.*

REMARK 5.9. *We may recall that the Well-Ordering Principle, which is equivalent to the Axiom of Choice in Mathematics or the so-called Zorn's lemma (which turned out to be falsely called as a lemma), is in fact a simple generalization to any set X of a property P satisfied by the set $\mathbb{N}$ of all natural numbers.*

The Well-Ordering Principle says that every set X has an ordering $>$ on it such that it satisfies the following property P: each non-empty subset S of X has a least element in S itself.

REMARK 5.10. *But, the main problem with this Principle is that the set (and consequently, any non-empty subset) of all natural numbers $\mathbb{N}$ is, to our best knowledge until now, the only known set equipped with an ordering $>$ on it satisfying such a property P, so that a "logical" generalization of such a very particular property of $\mathbb{N}$ to an arbitrary non-empty set X is quite difficult for us to reasonably accept it, although we nowadays accept it as an Axiom in Mathematics. For instance,*

EXAMPLE 5.13. *Even on the set $\mathbb{Z}$ of all integers, we have so far failed to figure out such an order $>$.*

REMARK 5.11. *But starting in the first quarter of the 20th century, mathematicians (mostly working in Real Analysis) first began to recognize the necessity of the Zorn's lemma in finishing the proof of some fundamental theorems, and then at around the middle of the 20th century it has been proved that this falsely called axiom, the Zorn's lemma, is in fact independent of any other axioms in mathematics and moreover, that the three are in fact all equivalent to each other.*

The application of this axiom is really needed in Algebra. For instance,

EXAMPLE 5.14. *We still do not know how to prove that every vector space over a field has a basis without using the Axiom of Choice.*

REMARK 5.12. *Another such an example where the application of the Zorn's lemma is definitely needed is the proof that each ideal in a commutative ring is contained in a maximal ideal (Krull's Theorem).*

REMARK 5.13. *The main problem with any one of these equivalent axioms in reality is the following: once it is used for establishment of an existence of a certain mathematical object, the method applied, nonetheless, will never ever tell us how to construct such an object at all, and consequently, we really must know when any one of them is actually applied in the proving process in order for us not to be mistakenly involved with an impossible attempt of constructing the claimed mathematical object. In other words, we believe that the axiom itself and all of its proved consequences are mathematically true, but this acknowledgement is nothing to do with the construction process of a certain mathematical object of which existence is guaranteed by such a logical, namely, non-computational, axiom.*

Proposition 5.4 allows us to have the following "**Second Principle of Mathematical Induction**, or **the Complete Induction**".

THEOREM 5.1. *If a subset $S \subseteq \mathbb{N}$ satisfies for each $n \in \mathbb{N}$ the following property,*

$$P(n): \qquad \text{if each } m < n \text{ is an element of } S, \text{ then } n \in S,$$

then $S = \mathbb{N}$.

PROOF. (By contradiction). Suppose that S is a proper subset of $\mathbb{N}$. Then $\mathbb{N} - S$ is a non-empty subset of $\mathbb{N}$; hence by Proposition 5.4, $\mathbb{N} - S$ contains a least element a; consequently, each $m < a$ is forced to be an element of S. That is, the hypothesis of $P(n)$ is satisfied. Therefore, $a \in S$, which is a contradiction. □

REMARK 5.14. *Because the set of all natural numbers $m < 1$ is the empty set $\varnothing$, the property $P(n)$ in this case reads "The assumption that* *every* $m \in \varnothing$ *belongs to* S *implies that* $1 \in S$*" and this is just the same as saying that "*$1 \in S$*". Thus, in applying the complete induction, we must establish the property* $P(n)$ *in particular for* $n = 1$*.*

We can also apply Peano's axioms to make recursive definition.

DEFINITION 5.14. *A* ***recursive definition*** *or* ***inductive definition*** *refers to the definition of a sequence of objects* C_n *indexed by the natural numbers in such a way that each* C_n *is defined in terms of the preceding one:*

(i) C_1 *is defined;*

(ii) *a rule is given for determining* C_{n+1} *from* C_n*.*

EXAMPLE 5.15. *Let* $C_n = x^n$*. A recursive definition of this function is*

$$x^1 = x \qquad \text{and} \qquad x^{n+1} = x \cdot x^n.$$

2.3. Addition and Multiplication of Integers.

NOTATION 5.2. *In this subsection, the elements of* $\mathbb{Z}$ *will be written as* a, b, c *while the elements of* $\mathbb{N}$ *are* k, m, n*, etc. Also in this subsection, our symbol* $\mathbb{N}$ *will denote the one with the additive identity "zero":*

$$\mathbb{N} := \mathbb{N} \cup \{0\},$$

where the $\mathbb{N}$ *in the right side is the usual one:* $\mathbb{N} = \{1, 2, 3, ...\}$ *without the additive identity "zero".*

We want to define addition and multiplication for integers, only using the corresponding operations for natural numbers.

DEFINITION 5.15. *Let* $d : \mathbb{N} \times \mathbb{N} \longrightarrow \mathbb{Z}$ *be a map defined by the rules:*

$$d(n + k, n) = k, \ d(n, n) = 0, \ d(m, m + k) = -k,$$

for each $m, n,$ *and* $k \neq 0$ *in* $\mathbb{N}$*,*

These rules do define the "*difference*" $d(m, n)$ for each pair (m, n), because of the law of trichotomy of the inequality in $\mathbb{N}$ (Proposition 5.1), and we sometimes simply call this function, the **difference function**.

REMARK 5.15. *The idea behind an introduction of the difference function for the pairs is that any integer can be written as a difference of two natural numbers:* $a = m - n$ *and* $b = k - l$ *in such a way that* $a + b$ *can be expressed as the difference* $(m + k) - (n + l)$*:*

$$a + b = (m - n) + (k - l) = (m + k) - (n + l),$$

under the "usual" addition of integers, which we do not formally defined yet.

Then, since we originally start with the set $\mathbb{Z}$ *of integers in defining addition and multiplication of integers only using the usual addition and multiplication of natural numbers, we need to associate* $\mathbb{Z}$ *to the Cartesian product* $\mathbb{N} \times \mathbb{N}$ *and then we come back via the difference function defined above.*

LEMMA 5.7. *The difference function* d *is surjective.*

PROOF. This is because, $k = d(k, 0)$ and $-k = d(0, k)$ for each $k \in \mathbb{N}$. □

This lemma suggests us how to associate $\mathbb{Z}$ to $\mathbb{N} \times \mathbb{N}$:

DEFINITION 5.16. *Let* $s : \mathbb{Z} \longrightarrow \mathbb{N} \times \mathbb{N}$ *be a map defined by the rules:*

$$s(m) = (m, 0), \;\; s(-k) = (0, k),$$

for $m \geq 0, k > 0$.

LEMMA 5.8. *The function* s *is a right inverse of the difference function* d.

PROOF. We note that

$$ds(m) = d((m, 0)) = m, \;\; d(s(-k)) = d((0, k)) = -k,$$

so ds is the identity map on $\mathbb{Z}$. □

REMARK 5.16. *Unfortunately,* s *is not a left inverse of* d*, because we have,*

$$sd((n + k, n)) = s(k) = (k, 0), \;\; sd((m, m + k)) = s(-k) = (0, k).$$

Now we work on the definitions and the immediate properties of the usual binary operations of addition and multiplication of two integers, using these two functions s and d:

DEFINITION 5.17. ***Addition***, $+$, *of pairs of natural numbers is defined as vector addition, i.e.,*

$$(m, n) + (m', n') = (m + m', n + n'),$$

for any $m, m', n, n' \in \mathbb{N}$.

LEMMA 5.9. *The binary operation, addition,* $+$ *on* $\mathbb{N} \times \mathbb{N}$ *is commutative, associative, and has the pair* $(0, 0)$ *as "the" additive identity.*

PROOF. Exercise. □

REMARK 5.17. *We note that*

$$\begin{aligned}(n, n) + sd((n + k, n)) &= (n + k, n),\\ (m, m) + sd((m, m + k)) &= (m, m + k).\end{aligned}$$

That is, to each pair u *there is another pair* u_0*, either* (n, n) *or* (m, m) *such that*

$$u = sd(u) + u_0, \ \ du_0 = 0. \tag{5.2}$$

Thus, the composition sd*, which is not the identity map on* $\mathbb{N} \times \mathbb{N}$*, missed being the identity map only by a pair* u_0 *of difference zero, i.e., a* diagonal *pair. Thus, if* u_0 *is any diagonal pair and* w *is any pair, then we have*

$$d(w + u_0) = dw.$$

Now, we are ready to define formally addition for integers in terms of the functions d and s:

DEFINITION 5.18. ***Addition*** $\oplus$ *of integers* a, b *is defined by*

$$a \oplus b = d(sa + sb).$$

LEMMA 5.10. (1) *If a and b are non-negative integers m and n, respectively, then*

$$a \oplus b = m + n.$$

(2) *If a and b are both negative integers $-m$ and $-n$, respectively, then*

$$a \oplus b = -(m + n).$$

(3) *If a is a non-negative integer m and b is a negative integer $-n$, then*

$$a \oplus b = d((m, n)) \ (= m - n).$$

PROOF. Part (1):

$$\begin{aligned} a \oplus b &= d(s(m) + s(n)) = d((m, 0) + (n, 0)) \\ &= d((m + n), 0) = m + n. \end{aligned}$$

Part (2):

$$\begin{aligned} a \oplus b &= d(s(-m) + s(-n)) = d((0, m) + (0, n)) \\ &= d((0, m + n)) = -(m + n). \end{aligned}$$

Part (3):

$$a \oplus b = d(s(m) + s(-n)) = d((m, 0) + (0, n)) = d((m, n)) = m - n.$$

□

LEMMA 5.11. *Taking the difference preserves addition:*

$$du \oplus dv = d(u + v),$$

for $u, v \in \mathbb{N} \times \mathbb{N}$.

PROOF. We write

$$u = sd(u) + u_0, \ \ v = sd(v) + v_0,$$

as in (5.2). Then

$$\begin{aligned} d(u+v) &= d(sdu+sdv+u_0+v_0) = d(sdu+sdv+u_0) \\ &= d(sdu+sdv) = du \oplus dv. \end{aligned}$$

□

COROLLARY 5.6. *For any $m, n \in \mathbb{N}$, we have:*

$$n \oplus d((m,n)) = m.$$

PROOF. Simply note that

$$\begin{aligned} n \oplus d((m,n)) &= d((n,0)) \oplus d((m,n)) = d((n,0)+(m,n)) \\ &= d((m+n,n)) = m. \end{aligned}$$

□

LEMMA 5.12. *Addition of integers, $a \oplus b$, is commutative, associative, and 0 is the additive identity.*

PROOF. By Lemma 5.11,

$$\begin{aligned} a \oplus b &= du \oplus dv = d(u+v) \\ &= d(v+u) = dv \oplus du = b \oplus a, \end{aligned}$$

where $a = du$ and $b = dv$ for some pairs u and v, because d is surjective, and hence it is commutative. Now,

$$\begin{aligned} a \oplus (b \oplus c) &= du \oplus (dv \oplus dw) = du \oplus (d(v+w)) \\ &= d(u+(v+w)) = d((u+v)+w) \\ &= d(u+v) \oplus d(w) = (du \oplus dv) \oplus d(w) \\ &= (a \oplus b) \oplus c, \end{aligned}$$

and hence it is associative. Since

$$a \oplus 0 = du \oplus d0 = d(u+0) = d(u) = a,$$

0 is the additive identity element. □

PROPOSITION 5.5. *$\mathbb{Z}$ has an associative and commutative binary operation $\oplus$ with 0 as the identity element and such that:*

(1) *(Subtraction is possible) For each $a, b \in \mathbb{Z}$, there exists $x \in \mathbb{Z}$ such that*

$$a \oplus x = b.$$

(2) *For any $m, n \in \mathbb{N}$,*

$$m + n = m \oplus n.$$

(3) *To each $a \in \mathbb{Z}$, there exist $m, n \in \mathbb{N}$ such that*

$$n \oplus a = m.$$

PROOF. We have already proved in the above lemma that it is associative and commutative.

Part (1): Write $a = d((m, n))$ and define $a' = d((n, m))$. Then since

$$(m, n) + (n, m) = (m + n, n + m) = w_0$$

is a diagonal pair, we have:

$$\begin{aligned} a \oplus a' &= d((m, n)) \oplus d((n, m)) \\ &= d((m, n) + (n, m)) = d(w_0) = 0. \end{aligned}$$

Thus, a' is an "additive inverse" of a. Now, for any $b \in \mathbb{Z}$, we have:

$$a \oplus (a' \oplus b) = (a \oplus a') \oplus b = 0 \oplus b = b,$$

thus the equation $a \oplus x = b$ for x has a solution $x = a' \oplus b$.

Part (2): We have already proved it in Lemma 5.10.

Part (3): Given $a \in \mathbb{Z}$, if a is a natural number n, then we have $0 \oplus n = 0 + n = n$. If a is $-p$ for some $p \geq 0$ then $p \oplus (-p) = 0$ by Part (1), and hence this proves Part (3). □

COROLLARY 5.7. *Subtraction is unique in $\mathbb{Z}$. That is, for any $a, b \in \mathbb{Z}$ there exists a unique $x \in \mathbb{Z}$ such that $a \oplus x = b$.*

PROOF. If $a \oplus x = a \oplus y$, we just add an inverse a' of a on both sides, to see:

$$\begin{aligned} a' \oplus (a \oplus x) &= (a' \oplus a) \oplus x = 0 \oplus x = x, \\ a' \oplus (a \oplus y) &= (a' \oplus a) \oplus y = 0 \oplus y = y. \end{aligned}$$

Thus, $x = y$. □

NOTATION 5.3. *From now on, we write $x = b - a$, and we stop using the special symbol $\oplus$ for addition in $\mathbb{Z}$.*

DEFINITION 5.19. ***Multiplication**, $\cdot$, of pairs of natural numbers in $\mathbb{N} \times \mathbb{N}$ is defined by the rule:*

$$(m, n) \cdot (m', n') = (mm' + nn', mn' + nm').$$

LEMMA 5.13. *This multiplication in $\mathbb{N} \times \mathbb{N}$ is commutative and associative with $(1, 0)$ as the identity element, and it is distributive over the vector addition of pairs defined in Definition 5.17.*

PROOF. Exercise. □

PROPOSITION 5.6. *$\mathbb{Z}$ has a commutative and associative binary operation, called multiplication, which is distributive over addition of Proposition 5.5 with 1 as the identity element, and such that the product of two natural numbers is the same in $\mathbb{N}$ as in $\mathbb{Z}$.*

PROOF. For any $a, b \in \mathbb{Z}$, define their product by

$$ab := d((sa)(sb)).$$

Then one can easily check this does define the usual multiplication in $\mathbb{Z}$. For example, if $a = m$ is non-negative and $b = -n$ is negative, then

$$m(-n) = d((m, 0)(0, n)) = d((0, mn)) = -mn.$$

Moreover, the difference function $d : \mathbb{N} \times \mathbb{N} \longrightarrow \mathbb{Z}$ satisfies

$$d(uv) = d(u)d(v),$$

for any $u = (m, n), v = (p, q) \in \mathbb{N} \times \mathbb{N}$, because, if $m \geq n$ and $p \geq q$ then since $du_0 = 0$ clearly implies $d(u_0 v) = 0$, we have by Lemma 5.13

$$\begin{aligned} d(uv) &= d((sd(u) + u_0)(sd(v) + v_0)) \\ &= d(sd(u)sd(v) + sd(u)v_0 + u_0 sd(v) + u_0 v_0) \\ &= d(sd(u)sd(v)) = d(sd(m, n)sd(p, q)) \\ &= d(s(m - n)s(p - q)) = d((m - n, 0)(p - q, 0)) \\ &= d((m - n)(p - q), 0) = (m - n)(p - q) \\ &= d(u)d(v), \end{aligned}$$

where u_o, v_o are diagonal pairs. Similarly, for other cases, we have: $d(uv) = d(u)d(v)$.

With this property and Lemma 5.11, one can verify that the distributive, associative, and other laws now carry over from $\mathbb{N} \times \mathbb{N}$ to $\mathbb{Z}$. □

PROPOSITION 5.7. *Sums and products of positive integers are positive, and for each $a \in \mathbb{Z}$, the trichotomy law holds: $a \in \mathbb{P}, a = 0$, or $-a \in \mathbb{P}$, where*

$$\mathbb{P} = \{1, 2, ...\},$$

is the set of positive *integers.*

PROOF. Exercise. □

DEFINITION 5.20. *In $\mathbb{Z}$, we define $a < b$ if there exists a positive m such that $a + m = b$.*

Then, the usual properties of "<" can be verified.

Now that $\mathbb{Z}$ with all the expected properties is at hand, there is no need to refer to the difference function $d : \mathbb{N} \times \mathbb{N} \longrightarrow \mathbb{Z}$ or its right inverse s, which are introduced and used in the construction above.

2.4. Polynomials in One Variable.

DEFINITION 5.21. *A **polynomial** with coefficients in a ring R is a linear combination of powers of the variable x:*

$$f(x) = a_n x^n + a_{n-1} x^{n-1} + \cdots + a_1 x + a_0, \tag{5.3}$$

*where $a_i \in R$, and such expressions are called **formal polynomials**, in order to distinguish them from* polynomial functions. *The set of all polynomials in one unknown with coefficients in R is denoted by*

$$R[x].$$

LEMMA 5.14. *Every formal polynomial with real coefficients determines a polynomial function on the real numbers.* □

The variable x appearing in (5.3) is an arbitrary symbol, that is, you can use any other symbol, say t, for the unknown, and the **monomials** x^i are considered *independent* in the following sense:

DEFINITION 5.22. *Two polynomials,*

$$\begin{aligned} f(x) &= a_n x^n + a_{n-1} x^{n-1} + \cdots + a_1 x + a_0, \\ g(x) &= b_m x^m + b_{m-1} x^{m-1} + \cdots + b_1 x + b_0, \end{aligned}$$

*are **equal** if they have the same coefficients, namely, $n = m$ and $a_i = b_i$ for $i = 0, ..., n$.*

REMARK 5.18. *We will see sometime later that because of this definition, the monomials x^i form a k-basis of the vector space, the set of all polynomials in one variable over a field k.*

DEFINITION 5.23. *Let $f(x) = a_n x^n + a_{n-1} x^{n-1} + \cdots + a_1 x + a_0$ be a non-zero polynomial.*

(1) *The **degree***

$$\deg(f(x))$$

of $f(x)$ is the largest integer k such that the coefficient a_k of $a_k x^k$ is non-zero.

(2) *The degree of the zero polynomial is considered to be indeterminate.*

(3) *The coefficient of highest degree of a polynomial which is not zero is called its **leading coefficient**.*

(4) *A **monic polynomial** is the one whose leading coefficient is* 1.

REMARK 5.19. *So far, since we did not formally construct an algebraic structure, called the polynomial ring $R[x]$ in one unknown x over a ring R, we can not have the equality:*

$$0x = 0.$$

On the other hand, the expression $0x$ actually occurs quite often if we "add" the same polynomial $3x$ twice with coefficients $3 = \bar{3} \in \mathbb{Z}/6\mathbb{Z}$,

$$3x + 3x = (3 + 3)x = 6x = 0x,$$

or, if we "multiply" two polynomials $2x$ and $3x$ with coefficients $2 = \bar{2}, 3 = \bar{3} \in \mathbb{Z}/6\mathbb{Z}$, then

$$2x \cdot 3x = (2 \cdot 3)x^2 = 0x^2.$$

Of course, since we want the set *$R[x]$ to be a commutative ring in which 0 is the identity element for addition of two polynomials, we must have $0x = 0$. This means that our construction process of making the set $R[x]$ into a ring must be good enough to justify that*

$$0x = 0.$$

Once this basic equality is established, $0x^2 = 0x \cdot x = 0x = 0$, etc., so that the "usual" addition and multiplication of two polynomials do make sense.

REMARK 5.20. *On the other hand, since any polynomial with real coefficients is a real-valued polynomial function:*

$$\begin{array}{c}\mathbb{R}^1\\ \downarrow\\ \mathbb{R},\end{array}$$

we know how to add and/or multiply such real functions in order to obtain other real-valued functions: for any real number $x \in \mathbb{R}$,

$$(f+g)(x) := f(x) + g(x) \qquad \text{and} \qquad (fg)(x) := f(x)g(x),$$

where x *is not a mere unknown symbol but is a variable having any real number as its value. Thus, if* $f(x) = 2x$ *and* $g(x) = 3x$*, then the product function* fg *has a real number as its value at a real coordinate* x*:*

$$(fg)(x) = 2x \cdot 3x = (2 \cdot 3) \cdot x^2 \in \mathbb{R}.$$

Also, if $f(x) = 3x$ *and* $g(x) = -3x$ *then their sum function* $f + g$ *has a real number as its value at a real coordinate* x*:*

$$(f+g)(x) = 3x + (-3)x = (3-3)x = 0x = 0 \in \mathbb{R}.$$

Here, the constant polynomial function 0 *plays the role of additive identity, and the constant polynomial function* 1 *plays the role of multiplicative identity, and consequently, "these real-valued polynomial functions,* $\mathbb{R}[x]$*", do form a* commutative ring. *In fact,* this ring is a subring of the ring of real-valued functions defined in Example 5.8.

REMARK 5.21. *Therefore, these two remarks do require us that our formal (i.e., abstract) construction of the ring* $R[x]$ *of polynomials in one unknown* x *must be done in such a way that its specialization of the coefficients* R*, to* $R = \mathbb{R}$*, does agree with the already existing "the ring* $\mathbb{R}[x]$ *of real-valued polynomial functions:* $\mathbb{R} \longrightarrow \mathbb{R}$*".*

2.5. Abstract Construction of a Polynomial Ring in One Unknown. Keeping these remarks in mind, we now introduce the following construction process for the ring of polynomials, which has already become a standard approach:

DEFINITION 5.24. *Let* R *be a commutative ring. Let* S *be the set of all infinite sequences* $(a_0, a_1, \ldots.)$*, where* $a_i \in R$*. That is,* S *is the collection of all maps:* $\mathbb{N} \cup \{0\} \longrightarrow R$*, and* T *a subset of* S *consisting of those sequences in* S *of which all but finitely many terms are zero.*

LEMMA 5.15. *There exists a natural one-to-one correspondence between the set $R[x]$ of polynomials with coefficients in R and the set T of those sequences: $\mathbb{N} \cup \{0\} \longrightarrow R$ of which all but finitely many terms, i.e., values, are zero.*

PROOF. Since polynomials are determined by their coefficients, the correspondence

$$a_0 + a_1x + a_2x^2 + ... + a_nx^n \longleftrightarrow (a_0, a_1, ..., a_n, 0, 0, ...),$$

where all the terms right after a_n are zero, is *natural, i.e., no other assignment is possible,* and indeed one-to-one. □

PROPOSITION 5.8. *Let R be a commutative ring. Then the set $R^{\mathbb{N}\cup\{0\}}$ of all infinite sequences: $\mathbb{N} \cup \{0\} \longrightarrow R$ becomes a commutative ring under the following operations:*

$$\begin{aligned} \textit{(vector addition)}\ \ (a_0, a_1, ...) + (b_0, b_1, ...) &= (a_0 + b_0, a_1 + b_1, ...), \\ \textit{(convolution product)}\ \ (a_0, a_1, ...) * (b_0, b_1, ...) &= (c_0, c_1, ...), \end{aligned}$$

where for each $k = 0, 1, ...$

$$c_k = \sum_{k=i+j} a_ib_j,$$

with the additive identity $(0, 0, ...)$ and the multiplicative identity $(1, 0, 0, ...)$ for the convolution product.

PROOF. It is straightforward to check that these sequences form an abelian group under vector addition. Since R is a commutative ring, so is the convolution product, while the special sequence $(1, 0, 0, ...)$ is served as a multiplicative identity.

To verify that the convolution product is associative, one simply expands the two expressions,

$$((a_0, a_1, ...) * (b_0, b_1, ...)) * (c_0, c_1, ...),$$
$$(a_0, a_1, ...) * ((b_0, b_1, ...) * (c_0, c_1, ...)),$$

separately to see that they are equal to the same sequence whose n-th term is $\sum_{n=i+j+k} a_i b_j c_k$, for $n = 0, 1, 2, ...$:

$$(..., \sum_{n=i+j+k} a_i b_j c_k, ...).$$

The distributive law is also similarly verified. □

COROLLARY 5.8. *The set T of those sequences: $\mathbb{N} \cup \{0\} \longrightarrow R$ of which all but finitely many terms, i.e., values, are zero, is a subring of the ring S.* □

COROLLARY 5.9. *The set $R[x]$ of polynomials with coefficients in a commutative ring R is a commutative ring containing R as a subring.*

PROOF. Since we have a one-to-one correspondence between $R[x]$ and T, and since T is a commutative ring, we define addition and multiplication of two polynomials exactly the same manner as those for sequences. Here, $0 \in R$ corresponds to the zero

$$(0, 0, 0, ...),$$

the $1 \in R$ corresponds to the one

$$(1, 0, 0, ...),$$

and x correspond to

$$(0, 1, 0, 0, ...).$$

Of course, $a \in R$ corresponds to

$$(a, 0, 0, ...),$$

and the set of sequences of the form $(a, 0, 0, ...)$ definitely forms a subring of T.

Thus, if we rewrite

$$\begin{aligned} 0 &= (0, 0, 0, ...), \\ 1 &= (1, 0, 0, ...), \\ x &= (0, 1, 0, 0, ...), \end{aligned}$$

then

$$\begin{aligned} x^2 &= (0,0,1,0,0,...), \\ x^3 &= (0,0,0,1,0,0,...), \\ &\cdots \end{aligned}$$

and so on; hence any element of T is expressed as a *polynomial* in x with coefficients in R:

$$\begin{aligned} (a_0, a_1, a_2, ..., a_n, 0, 0, ...) &= (a_0, 0, 0, ...) + (0, a_1, 0, 0, ...) \\ &\quad +(0, 0, a_2, 0, 0, ...) + \cdots \\ &= a_0 + a_1 x + a_2 x^2 + \cdots + a_n x^n, \end{aligned}$$

where the convolution product $*$ is replaced with the notation $\cdot$ for the "usual" multiplication of two polynomial expressions.

Therefore, these rewriting makes the commutative ring T into the set $R[x]$ with the "usual" addition and multiplication for polynomial expressions, making it into the "usual" polynomial ring, and R automatically becomes a subring of $R[x]$ under rewriting,

$$a = (a, 0, 0, ...),$$

for each $a \in R$. □

REMARK 5.22. *Thus, multiplication of* a *and* x *is well defined in* $R[x]$*, and hence in particular,*

$$0x = (0, 0, ...) * (0, 1, 0, 0, ...) = (0, 0, ...) = 0,$$

establishes

$$0x = 0$$

in $R[x]$*; consequently,*

$$0x^i = 0$$

for any $i = 0, 1, 2, ...,$ *and this fully justifies our usual way of writing a polynomial as a sum of* non-zero *monomials, and this finishes our abstract construction of the polynomial ring in one unknown* x *over a ring* R*.*

Thus, if

$$\begin{aligned} f(x) &= a_0 + a_1 x + a_2 x^2 + \cdots \\ g(x) &= b_0 + b_1 x + b_2 x^2 + \cdots \end{aligned}$$

are two polynomials, then

$$f(x) + g(x) = (a_0 + b_0) + (a_1 + b_1)x + (a_2 + b_2)x^2 + \cdots \tag{5.4}$$

is their vector addition or their *sum*, and

$$f(x)g(x) = \sum_{i,j} a_i b_j x^{i+j}. \tag{5.5}$$

is their convolution product or their *product.* If we collect those terms with the same degree, we can write their sum *in a standard form*:

$$f(x)g(x) = \sum_k c_k x^k,$$

where

$$c_k = \sum_{i+j=k} a_i b_j.$$

In these two formulas (5.4) and (5.5), we note that there are finitely many non-zero coefficients, $a_i + b_i$ and $a_i b_j$, respectively.

In summary:

PROPOSITION 5.9. *Let R be a commutative ring. Then there exists a* unique *commutative ring structure on the set of polynomials $R[x]$ having the following properties:*

(1) *Addition of polynomials is vector addition.*
(2) *Multiplication of polynomials is convolution product.*
(3) *The ring R is a subring of $R[x]$, when the elements of R are identified with the constant polynomials.*

PROOF. Because, once a set of polynomials in one unknown x with coefficients in a commutative ring R satisfies these properties, we can always identify it with the subring T of the ring S, as we have actually done above, and this is exactly what it means when we say that it has a unique commutative ring structure on the set of polynomials $R[x]$ having R as its subring. □

REMARK 5.23. *Now, we have "the" polynomial ring $R[x]$ at our hand. Since it is completely characterized by this proposition, we then can safely forget about how it has been constructed, and this is usually the case.*

2.6. Polynomials in Several Unknowns. Polynomials are fundamental to the theory of rings, and we must also consider polynomials, such as

$$x^2y^2 - 2x^2y + 3x^2 + 1,$$

in several unknowns. There is no major change in the definition.

DEFINITION 5.25. *Let $x_1, ..., x_n$ be $n \geq 1$ unknowns. A* ***monomial*** *is a formal product of these unknowns, of the form*

$$x_1^{e_1} x_2^{e_2} \cdots x_n^{e_n},$$

where the exponents e_i are non-negative integers.

We note that the n-tuple $(e_1, ..., e_n)$ of exponents determines the monomial, that is, there exists a one-to-one correspondence between the set of monomials and the set of such exponents. Such an n-tuple is called a **multi-index**.

NOTATION 5.4. *We write*

$$e = (e_1, ..., e_n)$$

for multi-indices. Using it, we may write the monomial symbolically as

$$x^e = x_1^{e_1} x_2^{e_2} \cdots x_n^{e_n},$$

where the monomial x^0, where $0 = (0, 0, ..., 0)$, is denoted by 1.

DEFINITION 5.26. *Let R be a commutative R. A* ***polynomial in several unknowns*** *with coefficients in a ring R is a finite linear combination of monomials, with coefficients in R. The product of a monomial* *by a non-zero element of R is also called a* ***monomial****. Thus,*

$$m = rx^e$$

is a monomial if $r \in R$ is non-zero and if $x^e = x_1^{e_1} x_2^{e_2} \cdots x_n^{e_n}$.

Using multi-index notation, formula (5.4) and formula (5.5) define addition and multiplication of polynomials in several unknowns, and the analogue of Proposition 5.9 holds.

NOTATION 5.5. *The ring of polynomials in several unknowns with coefficients in a commutative ring R is denoted by*

$$R[x_1, ..., x_n] \qquad \text{or} \qquad R[x],$$

where the symbol x is understood to refer to the ordered set of unknowns $(x_1, ..., x_n)$. When no set of unknowns has been introduced, $R[x]$ obviously refers to the polynomial ring in one unknown x.

Exercises for Chapter 5.2

EXERCISE 5.7. *Prove that the units of the polynomial ring $\mathbb{Q}[x]$ are the non-zero constant polynomials.*

EXERCISE 5.8. *Prove that every natural number n except 1 has the form m' for some natural number m.*

EXERCISE 5.9. *Prove that $\mathbb{Z}$ is an integral domain.*

3. Homomorphisms

DEFINITION 5.27. *Let R and R' be commutative rings.*

(1) *A* ***homomorphism*** *from a ring R to another ring R' or a* ***ring homomorphism****, $\varphi : R \longrightarrow R'$, is a map which is compatible with the laws of composition and which carries 1 to 1:*

$$\varphi(a + b) = \varphi(a) + \varphi(b), \qquad \varphi(ab) = \varphi(a)\varphi(b), \qquad \text{and} \qquad \varphi(1) = 1,$$

for each $a, b \in R$.

(2) *An **isomorphism** of rings is a bijective homomorphism.*

(3) *If there exists an isomorphism $R \longrightarrow R'$, then we say that R is **isomorphic** to R'.*

REMARK 5.24. *Thus, if $\varphi : R \longrightarrow R'$ is a ring homomorphism, then it is automatically a group homomorphism $R^+ \longrightarrow R'^+$, and it carries the additive identity 0 to the additive identity 0. But, since R is not a group with respect to multiplication $\cdot$, a ring homomorphism will not automatically carry the multiplicative identity 1 to the multiplicative identity; hence the condition $\varphi(1) = 1$ must be listed in the definition separately.*

EXAMPLE 5.16. *The zero map $R \longrightarrow R'$ is not a ring homomorphism unless R' is the zero ring. For, it does not send 1 to 1 unless $R' = \{0\}$, even though it is compatible with addition and multiplication.*

3.1. The Substitution Principle. Among the ring homomorphism, the most important ring homomorphisms are those obtained by *evaluating* polynomials at an element:

EXAMPLE 5.17. *Let $a \in \mathbb{R}$ be a real number. Then sending each real polynomial,*

$$f(x) \longmapsto f(a),$$

is clearly a ring homomorphism $\mathbb{R}[x] \longrightarrow \mathbb{R}$. If $i = \sqrt{-1}$, then sending a real polynomial,

$$f(x) \longmapsto f(i),$$

is clearly a ring homomorphism $\mathbb{R}[x] \longrightarrow \mathbb{C}$.

PROPOSITION 5.10. *(Substitution Principle) Let $\varphi : R \longrightarrow R'$ be a ring homomorphism. Then*

(1) *Given an element $\alpha \in R'$, there exists a* unique *ring homomorphism $\Phi : R[x] \longrightarrow R'$, sending*

$$x \longmapsto \alpha,$$

whose restriction to R is φ.

(2) *More generally, given elements* $\alpha_1, ..., \alpha_n \in R'$, *there exists a* unique *ring homomorphism* $\Phi : R[x_1, ..., x_n] \longrightarrow R'$, *sending*

$$x_i \longmapsto \alpha_i,$$

for each $i = 1, ..., n$, *whose restriction to* R *is* φ.

PROOF. Part (1). Suppose that there exists such a homomorphism Φ. For any polynomial

$$f(x) = a_0 + a_1x + \cdots + a_nx^n \in R[x],$$

$$\begin{aligned} \Phi(f(x)) &= \Phi(a_0) + \Phi(a_1x) + \cdots + \Phi(a_nx^n) \\ &= \varphi(a_0) + \Phi(a_1)\Phi(x) + \cdots + \Phi(a_n)\Phi(x^n) \\ &= \varphi(a_0) + \varphi(a_1)\Phi(x) + \cdots + \varphi(a_n)\Phi(x)^n \\ &= \varphi(a_0) + \varphi(a_1)\alpha + \cdots + \varphi(a_n)\alpha^n. \end{aligned} \tag{5.6}$$

Thus, such a homomorphism, if it exists, is uniquely determined by this formula, since whenever you are given two maps from the same domain to the same range the two maps are exactly the same if and only if their values at each point of the domain are exactly the same, and this is what's happening now.

Now, we note that such a homomorphism is conversely defined by this formula in (5.6): Given $\alpha \in R'$, the formula in (5.6) is clearly a homomorphism for addition and it sends 1 to 1; and if

$$g(x) = b_0 + b_1x + \cdots + b_mx^m$$

is another polynomial in $R[x]$, then their product

$$f(x)g(x) = \left(\sum_{i=0}^{n} a_ix^i\right)\left(\sum_{j=0}^{m} b_jx^j\right) = \sum_i \sum_j a_ib_jx^{i+j}$$

is mapped under Φ:

$$\begin{aligned}\Phi(f(x)g(x)) &= \sum_i \sum_j \varphi(a_i)\varphi(b_j)\alpha^{i+j} \\ &= \left(\sum_{i=0}^{n} \varphi(a_i)\alpha^i\right)\left(\sum_{j=0}^{m} \varphi(b_j)\alpha^j\right) = \Phi(f(x))\Phi(g(x)),\end{aligned}$$

by the distributive laws of both rings R and R', and hence it is a homomorphism for multiplication.

Since the proof of Part (2) is very similar to that of Part (1) if we adapt vector notation for indices, we omit its proof. □

Here is an application of the Substitution Principle in which the bottom ring R changes: Let $\psi : R \longrightarrow R_1$ be a ring homomorphism. Composing it with the canonical insertion $i : R_1 \hookrightarrow R_1[x]$ gives us a ring homomorphism:,

$$\varphi = i \circ \psi : R \longrightarrow R_1[x],$$

and hence by the Substitution Principle we have a unique ring homomorphism,

$$\Phi : R[x] \longrightarrow R_1[x],$$

which sends $x \longmapsto x$ and any element $a \in R$ to $\psi(a)$:

$$a_0 + a_1x + \cdots + a_nx^n \longmapsto \psi(a_0) + \psi(a_1)x + \cdots + \psi(a_n)x^n.$$

A typical case of this application is with the ring homomorphism $\mathbb{Z} \longrightarrow \mathbb{Z}_p = \mathbb{Z}/p\mathbb{Z}$, where p is a prime number so that $\mathbb{Z}_p$ is a field with p elements. By the Substitution Principle, this homomorphism is extended to a homomorphism

$$\mathbb{Z}[x] \longrightarrow \mathbb{Z}_p[x],$$

which sends

$$f(x) = a_0 + a_1x + \cdots + a_nx^n \longmapsto \bar{f}(x) = \bar{a}_0 + \bar{a}_1x + \cdots + \bar{a}_nx^n,$$

where $\bar{a}_i$ is the residue class of a_i modulo p. The polynomial $\bar{f}(x)$ is called the **residue of** $f(x)$ **modulo** p. Thus the kernel of this ring homomorphism is $(p) = p\mathbb{Z}[x]$.

Here is another example of the Substitution Principle in which the bottom field k is fixed: Let $k \to k$ be the identity map. Then for each $(a_1, ..., a_n) \in k^n$, we have the unique homomorphism

$$k[x_1, ..., x_n] \to k$$

sending each $x_i \mapsto a_i$, and this map is the so-called **evaluation homomorphism**

$$\varepsilon_{(a_1,...,a_n)} \equiv (a_1, ..., a_n)$$

at $X = (a_1, ..., a_n) \in k^n$. Moreover, if $\varphi : k[x_1, ..., x_n] \to k$ is the so-called a k-algebra homomorphism, then it is completely determined by its images $\varphi(x_i)$, $i = 1, 2, ..., n$. That is, we have shown that every homomorphism of k-algebras $k[x_1, ..., x_n] \to k$ is an evaluation homomorphism $\varepsilon_X \equiv X$ at some point $X = (a_1, ..., a_n) \in k^n$, and conversely.

The Substitution Principle also enables us to prove that various constructions of polynomial rings are equivalent:

COROLLARY 5.10. *Let $x = (x_1, ..., x_m)$ and $y = (y_1, ..., y_n)$ be two ordered-set of unknowns. Then there exists a unique isomorphism*

$$R[x, y] \longrightarrow R[x][y]$$

which send the variables to themselves and is the identity map on R.

PROOF. Since R is a subring of $R[x]$ and $R[x]$ is a subring of $R[x][y]$, R is also a subring of $R[x][y]$, and hence we have the inclusion map

$$\varphi : R \hookrightarrow R[x][y].$$

Thus by the Proposition 5.10 we have a unique ring homomorphism,

$$\Phi : R[x, y] \longrightarrow R[x][y],$$

which extends φ and sends the variables x_ν, y_ν to themselves.

To prove that Φ is an isomorphism, we construct its left and right inverse, again using Proposition 5.10: Since $R[x]$ is a subring of $R[x, y]$, we extend the inclusion map

$$\psi : R[x] \hookrightarrow R[x, y]$$

to a unique ring homomorphism,

$$\Psi : R[x][y] \longrightarrow R[x, y],$$

which sends each y_ν to itself and each $f(x) \in R[x]$ to itself. Thus, the composition

$$\Psi\Phi : R[x, y] \longrightarrow R[x, y]$$

is the identity map on R and it sends each x_ν, y_ν to themselves. On the other hand, given an inclusion map,

$$R \hookrightarrow R[x, y],$$

there exists only one ring homomorphism $R[x, y] \longrightarrow R[x, y]$ which is the identity map on R and which sends each x_ν, y_ν to themselves. But, since the identity map $R[x, y] \longrightarrow R[x, y]$ is already such a map, $\Psi\Phi$ must be the identity map. Similarly, we see that $\Phi\Psi$ is also the identity map. □

Here is an application of this corollary:

$$R[x, y] \approx R[x][y],$$

where the right side stands for the ring of polynomials in y whose coefficients are polynomials in x. The statement that these rings are isomorphic is a formalization of the procedure of collecting terms of like degree in y in a polynomial $f(x, y)$, in writing it as a polynomial in y.

EXAMPLE 5.18.

$$x^3y^3 - 2xy^2 + 3x^2y - 2y^2 + 5xy - 2 = (x^3)y^3 - (2(x+1))y^2 + (3x^2 + 5x)y - 2.$$

REMARK 5.25. *In our formal construction of the polynomial ring with more than one unknown over the ring R, say $R[x, y]$, the rewriting of a polynomial as is in this example was not a part of the construction. On the other hand, we can successively construct polynomial rings in several unknowns in an inductive manner:*

$$\begin{aligned} R &\hookrightarrow R[x] \hookrightarrow R[x][y] \hookrightarrow R[x][y][z] \hookrightarrow \cdots, \text{ or} \\ R &\hookrightarrow R[y] \hookrightarrow R[y][x] \hookrightarrow R[y][x][z] \hookrightarrow \cdots. \end{aligned}$$

Then one consequence of Proposition 5.10 is that we have, for instance, canonical isomorphisms:

$$R[x][y] \approx R[x,y] \approx R[y][x].$$

Here we need to recognize that the exact meaning of these isomorphisms is that the "rewriting process" *of polynomials as in the above example is an isomorphism of rings.*

3.2. Polynomial Functions. We note that any polynomial $f(x)$ with real coefficients is a polynomial function on the real line. We generalize this notion to

PROPOSITION 5.11. *Let $\mathcal{R}$ be the ring of continuous real-valued functions on $\mathbb{R}^n$:*

$$\mathcal{R} = \left\{ \begin{array}{c} \mathbb{R}^n \\ \downarrow \\ \mathbb{R} \end{array} \right\}.$$

Then the map $\varphi : \mathbb{R}[x_1, ..., x_n] \longrightarrow \mathcal{R}$ is an injective ring homomorphism.

PROOF. Since we clearly have an inclusion map $\mathbb{R} \hookrightarrow \mathcal{R}$, which sends each real number r to a real constant function $r((x_1, ..., x_n)) = r$, by Proposition 5.10, we have a unique extended ring homomorphism,

$$\varphi : \mathbb{R}[x_1, ..., x_n] \longrightarrow \mathcal{R},$$

which is identity on $\mathbb{R}$ and sends each unknown x_j to the variable x_j, namely, the coordinate function x_j.

Now, for the injectivity of φ, we suppose that we are given arbitrary two polynomials $f(x_1, ..., x_n)$ and $g(x_1, ..., x_n)$ such that the associated polynomial functions $f(x_1, ..., x_n)$ and $g(x_1, ..., x_n)$ are the same function $\mathbb{R}^n \longrightarrow \mathbb{R}$, so that their difference d as a polynomial function is the zero function.

If their difference d as a polynomial were not the zero polynomial, then some coefficient of d would not be zero; hence a suitable derivative of d would be a non-zero constant polynomial function; consequently the derivative of d, as a polynomial function, would not vanish at the origin, but this is impossible since their difference d as a polynomial function is by hypothesis the zero function. Thus, their difference d as a polynomial must be the zero polynomial, and this is what wanted to prove. $\square$

REMARK 5.26. *We note that this proof is based on the fact that the bottom field $\mathbb{R}$ is an* infinite *field. If the bottom field is a finite field, it may happen that the derivative of a non-zero polynomial function is the zero function. Thus, except for this problem, the term* polynomial *is often used to refer to a polynomial function associated to a polynomial.*

EXAMPLE 5.19. *Let $f(x) = x^2$ be a polynomial over $\mathbb{Z}_2 = \{0, 1\}$. Then its derivative is the zero function, because*

$$2 \cdot x^{2-1} = 0 \cdot x = 0.$$

On the other hand, the non-zero polynomial $g(x) = x^2 - x$ over the field $\{0, 1\}$ is clearly the zero function.

3.3. The Unique Structural Homomorphism from $\mathbb{Z}$.

PROPOSITION 5.12. *There exists exactly one ring homomorphism from $\mathbb{Z}$ to any commutative ring R,*

$$\varphi : \mathbb{Z} \longrightarrow R,$$

which is defined by

$$\varphi(n) = \begin{cases} \text{“}n \textit{ times } 1_R\text{”} = 1_R + \cdots + 1_R \textit{ if } n \geq 0, \\ -\varphi(-n) \textit{ if } n < 0. \end{cases}$$

PROOF. (Uniqueness) Let $\varphi : \mathbb{Z} \longrightarrow R$ be any ring homomorphism. Since it is a ring homomorphism, it satisfies:

$$\begin{aligned} \varphi(1) &= 1, \\ \varphi(n+1) &= \varphi(n) + \varphi(1). \end{aligned}$$

Thus,

$$\varphi(n+1) = \varphi(n) + 1.$$

But this just means that φ is in fact determined on the natural numbers by the following recursive definition:

$$\varphi(1) = 1, \qquad \text{and} \qquad \varphi(\sigma(n)) = \varphi(n) + 1,$$

where σ denotes the successor function in Definition 5.7. Since any ring homomorphism is a morphism of abelian groups, we know that φ additionally satisfies:

$$\begin{aligned} \varphi(-n) &= -\varphi(n) \qquad \text{if } n > 0, \\ \varphi(0) &= 0. \end{aligned}$$

Therefore, any ring homomorphism $\varphi : \mathbb{Z} \longrightarrow R$ satisfies the formula in the definition of the proposition; consequently the map defined by the formula in the proposition is the only possible one.

(Existence) The definition in the proposition clearly defines a map $\mathbb{Z} \longrightarrow R$. By induction on n we are going to prove that this φ is a ring homomorphism. Let m be an arbitrary positive integer.

If $n = 1$, then

$$\varphi(m+1) = \varphi(m) + 1 = \varphi(m) + \varphi(1).$$

We thus assume, as an induction hypothesis, that

$$\varphi(m+n) = \varphi(m) + \varphi(n),$$

hold for an arbitrary positive integer m and some positive integer $n > 1$. Then

$$\begin{aligned} \varphi(m+\sigma(n)) &= \varphi(m+(n+1)) \\ &= \varphi((m+n)+1) \qquad \text{(properties of addition of integers)} \\ &= \varphi(m+n)+1 \qquad \text{(definition of } \varphi) \\ &= \varphi(m)+\varphi(n)+1 \qquad \text{(induction hypothesis)} \\ &= \varphi(m)+\varphi(n+1) \qquad \text{(definition of } \varphi) \\ &= \varphi(m)+\varphi(\sigma(n)), \end{aligned}$$

and hence by the Induction Principle we conclude that

$$\varphi(m+n) = \varphi(m) + \varphi(n),$$

for every *positive* integers m and n.

Now if both $m < 0$ and $n < 0$ then $m = -a$ and $n = -b$ with positive integers a, b. Thus,

$$\begin{aligned} \varphi(m+n) &= \varphi(-(a+b)) = -\varphi(a+b) \\ &= -(\varphi(a) + \varphi(b)) = -\varphi(a) - \varphi(b) \\ &= \varphi(-a) + \varphi(-b) = \varphi(m) + \varphi(n). \end{aligned}$$

Now if $m < 0$, but $n > 0$ *and* $m + n > 0$, then $m = -a$ with $a > 0$. Thus,

$$\begin{aligned} \varphi(m+n) &= \varphi(-a+n) = \varphi(n-a) \\ &= (n-a) \cdot 1_R \\ &= \underbrace{1_R + \cdots + 1_R}_{n \text{ times}} \underbrace{-1_R - \cdots - 1_R}_{a \text{ times}} \\ &= n \cdot 1_R - a \cdot 1_R \\ &= \varphi(n) - \varphi(a) = \varphi(n) + \varphi(-a) \\ &= \varphi(n) + \varphi(m) = \varphi(m) + \varphi(n). \end{aligned}$$

Now if $m < 0$, but $n > 0$ *and* $m + n < 0$, then $m = -a$, $m + n = -b$ with $a > 0$, $b > 0$. Thus, $b = a - n$ and

$$\begin{aligned} \varphi(m+n) &= -\varphi(-(m+n)) = -(-(m+n) \cdot 1_R) \\ &= -(a-n) \cdot 1_R = -(a \cdot 1_R - n \cdot 1_R) \\ &= n \cdot 1_R - a \cdot 1_R = \varphi(-a) + \varphi(n) \\ &= \varphi(m) + \varphi(n). \end{aligned}$$

Now if $m > 0, n < 0$, but $m + n = 0$, then $n = -m$ and

$$\varphi(m+n) = 0 = \varphi(m) - \varphi(m) = \varphi(m) + \varphi(n).$$

Thus this proves, that

$$\varphi(m+n) = \varphi(m) + \varphi(n),$$

for *every* integers m and n.

Now if $n = 1$, then

$$\begin{aligned}\varphi(m \cdot 1) &= \varphi(m)\\ &= \varphi(m) \cdot 1\\ &= \varphi(m)\varphi(1).\end{aligned}$$

We thus assume, as an induction hypothesis, that

$$\varphi(mn) = \varphi(m)\varphi(n),$$

holds for an arbitrary positive integer m and some positive integer $n > 0$. Then

$$\begin{aligned}\varphi(m \cdot \sigma(n)) &= \varphi(m(n+1)) = \varphi(mn + m)\\ &= \varphi(mn) + \varphi(m) = \varphi(m)\varphi(n) + \varphi(m)\\ &= \varphi(m)(\varphi(n) + 1) = \varphi(m)\varphi(n+1)\\ &= \varphi(m)\varphi(\sigma(n)),\end{aligned}$$

and hence by the Induction Principle we conclude that

$$\varphi(mn) = \varphi(m)\varphi(n),$$

for every *positive* integers m and n.

Now if $m < 0$ and $n > 0$ then $m = -a$ with $a > 0$; hence

$$\begin{aligned}\varphi(mn) &= \varphi(-an) = -\varphi(an)\\ &= -\varphi(a)\varphi(n) = \varphi(-a)\varphi(n)\\ &= \varphi(m)\varphi(n).\end{aligned}$$

Now if $m < 0, n < 0$, then $m = -a$, $n = -b$ with $a > 0$, $b > 0$. Thus,

$$\begin{aligned}\varphi(mn) &= \varphi((-a)(-b)) = \varphi(ab)\\ &= \varphi(a)\varphi(b) = -\varphi(m) \cdot -\varphi(n)\\ &= \varphi(m)\varphi(n).\end{aligned}$$

Moreover, we have:

$$\varphi(m \cdot 0) = \varphi(0) = 0 = \varphi(m)\varphi(0).$$

This proves that

$$\varphi(mn) = \varphi(m)\varphi(n),$$

for *every* integers m and n. □

EXAMPLE 5.20. *We can interpret the* symbol

$$3$$

as the element $1+1+1$ *in* R, *and hence we can interpret an integer polynomial such as*

$$3x^3 + 2x$$

as an element of the polynomial ring $R[x]$.

Exercises for Chapter 5.3

EXERCISE 5.10. *Prove that there exists only one ring homomorphism* $\mathbb{Z}_{12} \longrightarrow \mathbb{Z}_3$.

EXERCISE 5.11. *Prove that* $\mathbb{Z}_6$ *is isomorphic to* $\mathbb{Z}_2 \times \mathbb{Z}_3$.

EXERCISE 5.12. *Prove that there can be no ring homomorphism* $\mathbb{Z}_3 \longrightarrow \mathbb{Z}_4$.

EXERCISE 5.13. *For* m *and* n *positive integers, find condition that there be a ring homomorphism* $\mathbb{Z}_m \longrightarrow \mathbb{Z}_n$.

EXERCISE 5.14. *Prove that the inverse of a ring isomorphism is an isomorphism.*

EXERCISE 5.15. *Prove that the group of all automorphisms of the polynomial ring* $F[x]$ *over a field* F, which leaves all the constant polynomials, i.e., those elements in F, fixed, *consists of substitutions given by*

$$x \longmapsto ax + b, \qquad a \neq 0, b \in F.$$

4. Ideals

Let $\varphi : R \longrightarrow R'$ be a ring homomorphism. Since it is automatically a group homomorphism, it has its **kernel**,

$$\ker \varphi = \varphi^{-1}(0),$$

and we have already learned that $\ker \varphi$ is a subgroup of R^+ under the law of addition; hence it is automatically normal subgroup of R^+. Moreover, we have also learned that the kernel of a group homomorphism is a subgroup of the domain, which is moreover a normal subgroup of the domain.

LEMMA 5.16. *The kernel of a ring homomorphism $\varphi : R \longrightarrow R'$ is closed under addition and multiplication, and it satisfies the "**absorption property**":*

If $a \in \ker \varphi$ and $r \in R$, then $ra \in \ker \varphi$.

PROOF. If $a, b \in \ker \varphi$, then $\varphi(a + b) = \varphi(a) + \varphi(b) = 0 + 0 = 0$, $\varphi(ab) = \varphi(a)\varphi(b) = 0$, and hence $a + b \in \ker \varphi$ and $ab \in \ker \varphi$. Now, $\varphi(ra) = \varphi(r)\varphi(a) = \varphi(r)0 = 0$ if $a \in \ker \varphi$. □

REMARK 5.27. *But, the identity element of R is not contained in $\ker \varphi$, unless $\ker \varphi = R$. Moreover, if $\ker \varphi = R$, then $1 \in R$ is mapped to $0 \in R'$, which is possible only when the range R' is forced to be the zero ring. That is, if a given R' is not the zero ring, then $\ker \varphi$ is a proper subset of R.*

This gives rise to the following definition:

DEFINITION 5.28. *An **ideal** I of a commutative ring R is a subset of R satisfying the following properties:*

(1) *I is a subgroup of R^+;*
(2) *(Absorption Property) If $a \in I$ and $r \in R$, then $ra \in I$.*

EXAMPLE 5.21. *The kernel of any ring homomorphism is an ideal, naturally associated to the homomorphism. For instance, the kernel of the evaluation homomorphism at the real number 2:* $\varphi : \mathbb{R}[x] \longrightarrow \mathbb{R}$ *consists of those real polynomials which are multiples of the linear polynomial* $x-2$.

LEMMA 5.17. *Let* I *be a subset of* R. *Then the following statements are equivalent:*

(1) I *is an ideal of* R.
(2) I *is a non-empty subset of* R *and for any* $a_i \in I$ *and any* $r_i \in R$, *their finite linear combination,*

$$\sum_{i=1}^{n} r_i a_i = r_1 a_1 + \cdots + r_n a_n,$$

is an element of I.

PROOF. Exercise. □

4.1. Principal Ideals.

DEFINITION 5.29. *Let* $a \in R$ *be an element of a commutative ring* R. *The subset,*

$$(a) = aR = \{ra : r \in R\},$$

is called the ***principal ideal*** *generated by* a.

EXAMPLE 5.22. *The kernel of the evaluation homomorphism at a real number* $a \in \mathbb{R}$, $\varphi : \mathbb{R}[x] \longrightarrow \mathbb{R}$ *is the principal ideal generated by the polynomial* $x - a$,

$$(x-a)\mathbb{R}[x].$$

LEMMA 5.18. aR *is an ideal of* R.

PROOF. Exercise. □

Obviously, it is the smallest ideal of R containing the set $\{a\}$.

DEFINITION 5.30. *Let* $a_1, ..., a_n \in R$ *be any* n *elements of* R. *The subset*

$$(a_1, ..., a_n) = \{\sum_{i=1}^{n} r_i a_i : r_i \in R\} \tag{5.7}$$

is called the ***ideal generated by*** $a_1, ..., a_n$.

EXAMPLE 5.23. *Let* $R = \mathbb{Z}[x]$. *Then* $(2, x)$ *denotes the ideal generated by* 2 *and* x. *Thus, it consists of those integer polynomials whose constant term is divisible by* 2. *Thus, this ideal is in fact the kernel of the homomorphism* $\mathbb{Z}[x] \longrightarrow \mathbb{Z}/2\mathbb{Z}$ *which sends*

$$f(x) \longmapsto \textit{residue of } f(0) \bmod(2).$$

DEFINITION 5.31. *Let* R *be a commutative ring.*

(1) *The set* $I = (0)$ *is called the* ***zero ideal****.*
(2) *The ideal* $(1) = R$ *is called the* ***unit ideal****.*

An ideal I *of* R *is called* ***proper*** *if it is neither* (0) *nor* R.

PROPOSITION 5.13. *A commutative ring* R *is a field if and only if it has only two ideals,* (0) *and* (1).

PROOF. Since every non-zero element of a field is invertible, the $\Longrightarrow$ direction holds. Conversely, suppose that R has only two ideals. Since the "zero" ring, $R = \{0\}$, has only one ideal, our ring R under consideration is not the zero ring. Thus the only two ideals are necessarily (0) and (1), because these two are ideals of any ring R. To prove that every non-zero element of R is invertible, let a be a non-zero element of R. Then the principal ideal (a) generated by a is necessarily equal to (1) and hence a multiple of a is 1. That is, a is invertible. □

COROLLARY 5.11. *Every ring homomorphism* φ *from a field* F *to a non-zero ring* R *is injective.*

PROOF. If $\ker\varphi = \{1\}$, then everything is mapped to the zero, and hence φ is the zero map. But, this is impossible because any ring homomorphism sends 1 to 1. Thus, $\ker\varphi$ must be (0). □

PROPOSITION 5.14. *Every ideal in the ring* $\mathbb{Z}$ *of integers is principal.*

PROOF. This is exactly what we have proved in Proposition 2.3 of Chapter Two. □

In particular, this proposition enables us to introduce a new concept, called, the characteristic of a ring R.

DEFINITION 5.32. *Let R be a commutative ring. The* ***characteristic*** *of R is the single generator of the kernel of the unique structural ring homomorphism $\varphi : \mathbb{Z} \longrightarrow R$.*

Thus, by abuse of language, the characteristic of R is the smallest non-negative integer n such that

$$n \cdot 1_R = 1 + \cdots + 1 \text{ (n times) } = 0.$$

EXAMPLE 5.24. *$\mathbb{Z}, \mathbb{Q}, \mathbb{R}$, and $\mathbb{C}$ have characteristic zero, but for any prime number p the field $\mathbb{Z}/p\mathbb{Z}$ with p elements has characteristic p.*

The proof that every ideal of the ring $\mathbb{Z}$ of integers is principal can be adapted to show that every ideal in the polynomial ring $F[x]$ is principal. To prove this, we need division with remainder for polynomials.

PROPOSITION 5.15. *Let $f(x), g(x)$ be two polynomials in one variable with coefficients in a ring R. Assume that the leading coefficient of $f(x)$ is a* unit, i.e., an invertible element, *in R. Then there exist* unique *polynomials $q(x), r(x) \in R[x]$ such that*

$$g(x) = q(x)f(x) + r(x),$$

where the degree of the remainder $r(x)$ is (strictly) less than the degree of $f(x)$ or else $r(x) \equiv 0$.

PROOF. (Existence: By induction on the degree of $g(x)$.) If $\deg(g(x)) < \deg(f(x))$, then simply take $q(x) = 0$; hence $r(x) = g(x)$, in this case. Otherwise, we write

$$\begin{aligned} f(x) &= a_m x^m + \cdots + a_1 x + a_0, \\ g(x) &= b_n x^n + \cdots + b_1 x + b_0, \end{aligned}$$

where $m \leq n$, $a_i, b_j \in R$, both the leading coefficients a_m and b_n are non-zero, and a_m is a unit in R. If $n = 0$, then $m = 0$, $g(x) = b_0$, and $f(x) = a_0$ a unit. Thus,

$$q(x) = b_0 a_0^{-1} \quad \text{and} \quad r(x) = 0,$$

Then $g(x) = b_0 = (b_0 a_0^{-1})a_0 = q(x)f(x) + r(x)$. Now assume that the existence of such an expression in the proposition is true for polynomials of degree less than $n = \deg(g(x))$. Then since

$$(b_n a_m^{-1} x^{n-m})f(x)$$

has degree n and leading coefficient b_n, we see that

$$g(x) - (b_n a_m^{-1} x^{n-m})f(x) = b_n x^n + \cdots + b_0 - (b_n x^n + \cdots + b_n a_m^{-1} a_0 x^{n-m})$$

is certainly a polynomial of degree less than n. Thus, by induction hypothesis there exist polynomials $q_1(x)$ and $r(x)$ such that

$$g(x) - (b_n a_m^{-1} x^{n-m})f(x) = q_1(x)f(x) + r(x),$$

where either $\deg(r(x)) < \deg(f(x))$ or $r(x) \equiv 0$. Thus, we have

$$g(x) = q(x)f(x) + r(x),$$

where $q(x) = b_n a_m^{-1} x^{n-m} + q_1(x)$.

(Uniqueness) Suppose we have another such expression:

$$g(x) = q_1(x)f(x) + r_1(x).$$

Then their difference is,

$$(q(x) - q_1(x)) \cdot f(x) = r_1(x) - r(x).$$

We then note that the leading coefficient a_m is a unit so that in particular it is not a zero-divisor; hence if $q(x)$ and $q_1(x)$ were *not* the same polynomial, then we would have

$$\begin{aligned} \deg(r_1(x) - r(x)) &= \deg((q(x) - q_1(x))f(x)) \\ &= \deg(q(x) - q_1(x)) + \deg(f(x)) \\ &\geq \deg(f(x)), \end{aligned}$$

but is impossible to happen, because we already know that

$$\deg(r_1(x) - r(x)) \leq \max(\deg(r_1(x)), \deg(r(x))) < \deg(f(x)).$$

Therefore, $q(x) - q_1(x) = 0$ and consequently we also have $r(x) - r_1(x) = 0$. □

REMARK 5.28. *We note that if the coefficient ring R is a field, then the assumption that the leading coefficient a_m of $f(x)$ is a unit, i.e., an invertible element, is satisfied, provided that there is a leading coefficient, that is, that $f(x) \neq 0$.*

COROLLARY 5.12. *Let $g(x)$ be a polynomial in $R[x]$. Let α be an element of R. Then $g(\alpha) = 0$ if and only if $x - \alpha$ divides $g(x)$ in $R[x]$.*

PROOF. Apply the proposition with $f(x) = x - \alpha \in R[x]$. □

PROPOSITION 5.16. *Let F be a field. Then the polynomial ring $F[x]$ in one variable x over F is a Principal Ideal Domain (PID).*

PROOF. One can easily prove that $F[x]$ is an integral domain. Since the zero ideal (0) is certainly generated by the element 0, we may assume that a proper ideal I of $F[x]$ contains a non-zero element d. If I contains a non-zero constant polynomial d in the bottom field F, then $1 = d^{-1}d \in I$; hence $I = (1) = F[x]$. Otherwise, we choose d to be a non-zero non-constant polynomial in I of minimum degree. We claim that I is the principal ideal generated by d. Certainly, by definition, $(d) \subseteq I$. To prove the reverse direction, we go as follows: For any polynomial $f \in I$, we have

$$f = ad + r$$

for uniquely determined quotient polynomial a and the remainder polynomial r such that $0 \leq \deg(r) < \deg(d)$. Then since

$$r = f - ad \in I,$$

and since $\deg(r) < \deg(d)$, we must have $r \equiv 0$ (i.e., the zero polynomial) because of our choice of the polynomial d, having minimal degree among all of non-zero non-constant polynomials in I; thus $f = ad$, showing that $I \subseteq (d)$. □

DEFINITION 5.33. *A **principal ideal domain** (PID) is an integral domain where every ideal is generated by a single element in it.*

COROLLARY 5.13. *Let* $f(x), g(x)$ *be two* non-zero *polynomials in one variable* x *over a field* F. *Then there exists a unique monic polynomial* $d(x)$, *called* ***the greatest common divisor*** *of* $f(x)$ *and* $g(x)$, *satisfying:*

(1) $d(x)$ *divides both* $f(x)$ *and* $g(x)$.
(2) *If* $h(x)$ *divides both* $f(x)$ *and* $g(x)$, *then* $h(x)$ *divides* $d(x)$.
(3) $d(x)$ *generates the ideal* $(f(x), g(x))$ *of* $F[x]$ *generated by the two polynomials* $f(x)$ *and* $g(x)$.
(4) *There exist (non-necessarily uniquely determined) two polynomials* $p(x)$ *and* $q(x)$ *such that*

$$d(x) = p(x)f(x) + q(x)g(x).$$

PROOF. The proof of this corollary is similar to that of Proposition 2.5 of Chapter Two and its corollary. □

Exercises for Chapter 5.4

EXERCISE 5.16. *Display the inclusion diagram of all ideals of the ring* $\mathbb{Z}_{48}$.

EXERCISE 5.17. *Prove that the kernel of a ring homomorphism is an ideal.*

EXERCISE 5.18. *Let* $\varphi : R \longrightarrow S$ *be a ring homomorphism,* I *an ideal of* S. *Prove that the inverse image* $\varphi^{-1}(I)$ *of* I *is an ideal of* R.

EXERCISE 5.19. *Prove that if an ideal contains a unit, then* $I = (1)$, *the unit ideal.*

EXERCISE 5.20. *Prove that every non-zero ideal in* $\mathbb{Z}[i]$ *contains a non-zero integer.*

EXERCISE 5.21. *Show that the set of all integer polynomials with even constant term is an ideal which is not a principal ideal.*

EXERCISE 5.22. *Prove that each ideal in* $\mathbb{Z}_n$ *is principal.*

EXERCISE 5.23. *Find a non-principal ideal in* $\mathbb{Z}_9[x]$.

EXERCISE 5.24. *Determine the kernel of the map* $\mathbb{Z}[x] \longrightarrow \mathbb{R}$ *sending* $f(x) \longmapsto f(1+\sqrt{3})$.

EXERCISE 5.25. *Describe the kernel of the map* $\mathbb{C}[x,y,z] \longrightarrow \mathbb{C}[t]$ *sending*

$$\begin{aligned} x &\longmapsto t \\ y &\longmapsto t^2 \\ z &\longmapsto t^3. \end{aligned}$$

EXERCISE 5.26. *Let* I, J *be two ideals in a ring* R. *Prove that* $I \cap J$ *is an ideal of* R.

EXERCISE 5.27. *Prove that* $(2) \cap (x) = (2x)$ *in* $\mathbb{Z}[x]$.

EXERCISE 5.28. *Let* I, J *be two ideals in a ring* R.

(1) *Show by example that* $I \cup J$ *need not be an ideal.*
(2) *Prove that*

$$I + J = \{r \in R : r = i + j \text{ for some } i \in I, j \in J\}$$

is an ideal of R. *This ideal is called the* ***sum*** *of the ideals* I *and* J.

(3) *Show by example that the set of products*

$$\{ij : i \in I, j \in J\}$$

need not be an ideal. [Hint: Consider $(2,x)$, $(3,x)$ *of* $\mathbb{Z}[x]$. *and note that* $x = 3x - 2x$ *cannot be a product* $\alpha\beta$ *of any elements* $\alpha \in (2,x)$ *and* $\beta \in (3,x)$.*]*

(4) *Prove that the set of finite sums*

$$IJ := \left\{ \sum_{\nu=1}^{n} x_\nu y_\nu : x_\nu \in I, y_\nu \in J \right\}$$

is an ideal. This ideal is called the ***product*** *of the ideals* I *and* J.

EXERCISE 5.29. *Let* I, J *be two ideals in a ring* R.

(1) *Prove that*

$$IJ \subset I \cap J.$$

(2) *Show by example that IJ and $I \cap J$ need not be the same.*

EXERCISE 5.30. *Find an ideal in the polynomial ring $F[x, y]$ in two unknowns over a field F which is not a principal ideal.*

EXERCISE 5.31. *Prove that the characteristic of a field is either zero or a prime integer.*

EXERCISE 5.32. *Determine all automorphisms of the ring $\mathbb{Z}[x]$.*

5. Quotient Rings

Let I be an ideal of a ring R. Since I has the additive subgroup of the additive group R^+, the cosets of I are the subsets

$$a + I, \qquad a \in R,$$

and from the theory developed in the Chapter called Groups, we know that the set of cosets $\bar{R} = R/I$ is a group under the induced addition. Moreover, $\bar{R}$ has a multiplication naturally induced from that of R, and consequently $\bar{R}$ becomes a ring:

THEOREM 5.2. *Let I be an ideal of a ring R.*

(1) *There exists a unique ring structure on the set of cosets $\bar{R} = R/I$ such that the canonical map $\pi : R \longrightarrow \bar{R}$ which sends*

$$a \longmapsto a + I$$

is a ring homomorphism.

(2) *The kernel of the ring homomorphism π is I.*

PROOF. Part (1). We have proved it when $R = \mathbb{Z}$, the ring of integers in Chapter Two. In putting a ring structure on $\bar{R}$ we have already seen that it has the induced law of addition from the law of addition of R in Chapter Two, so that we only need to define multiplication on $\bar{R}$ which is also going to be naturally induced from the law of multiplication on

R. For this, let $x, y \in \bar{R}$. Then $x = a + I$ and $y = b + I$ for some (non-uniquely chosen) $a, b \in R$. Then for any elements $r \in a + I, s \in b + I$, we have $r = a + p, s = b + q$ for some elements $p, q \in I$; hence their product

$$rs = (a + p)(b + q) = ab + aq + pb + pq,$$

where $aq + pb + pq \in I$, since I is an ideal of R. Thus, $rs \in ab + I$. Similarly, if $r \in a' + I$ and $s \in b' + I$ for some other $a', b' \in R$, still with $a + I = a' + I$ and $b + I = b' + I$, then $rs \in a'b' + I$. This suggests us to define their product xy by

$$xy := ab + I.$$

We need to verify that this definition is **well-defined**: Let $x = a' + I$ and $y = b' + I$ for some other $a', b' \in R$. We need to show that $ab + I \subseteq a'b' + I$ (and the reverse inclusion), but this is easy: if $\alpha \in ab + I$, then $\alpha = ab + p_1$ for some $p_1 \in I$. Thus, if

$$\begin{aligned} r &= a + p = a' + p', \\ s &= b + q = b' + q', \end{aligned}$$

for some $p, p', q, q' \in I$, then the product

$$\begin{aligned} rs &= (a + p)(b + q) = ab + aq + bp + pq \\ &= (a' + p')(b' + q') = a'b' + a'q' + b'p' + p'q', \end{aligned}$$

says that

$$ab = a'b' + a'q' + b'p' + p'q' - (aq + bp + pq).$$

Thus,

$$\begin{aligned} \alpha &= ab + p_1 \\ &= a'b' + a'q' + b'p' + p'q' - (aq + bp + pq) + p_1 \\ &= a'b' + p_1', \end{aligned}$$

with $p_1' = a'q' + b'p' + p'q' - (aq + bp + pq) + p_1 \in I$. This proves that $\alpha \in a'b' + I$, establishing

$$ab + I \subseteq a'b' + I.$$

Then changing the role of a, b with a', b', we also see that

$$a'b' + I \subseteq ab + I.$$

In this way, we have verified in detail that

$$\begin{aligned}(a+I)(b+I) &= ab + I \\ &= a'b' + I \\ &= (a'+I)(b'+I)\end{aligned}$$

where a, a' represent the same one coset and b, b' represent the same another coset. Therefore, the product defined above is well-defined, that is, it is not dependent on the choice of the representatives of cosets of I.

Then those defining axioms for $\bar{R}$ to be a commutative ring are automatically satisfied. We leave the details of the verification as an exercise.

Moreover, sending $a \longmapsto a + I$ is clearly a ring homomorphism whose kernel is I: For this, we simply note that $a + I = b + I$ if and only if $a - b \in I$, because $a \in \pi^{-1}(0 + I)$ if and only if $\pi(a) = a + I = 0 + I$ if and only if $a - 0 \in I$. □

THEOREM 5.3. *(Mapping property of quotient rings) Let $f : R \longrightarrow R'$ be a ring homomorphism with kernel I and let J be an ideal of R such that $J \subseteq I$. If we denote the quotient ring R/J by $\bar{R}$, the following statements are true:*

(1) *There exists a unique ring homomorphism $\bar{f} : \bar{R} \longrightarrow R'$ such that $\bar{f} \circ \pi = f$ with kernel I/J:*

$$\begin{array}{ccc} R & \xrightarrow{\pi} & \bar{R} = R/J \\ & \searrow & \downarrow \bar{f} \\ & & R' \end{array} .$$

(2) *(First Isomorphism Theorem) If* $J = I$*, then* $\bar{f}$ *maps isomorphically to the image of* f*.*

PROOF. Part (1). We define $\bar{f}$ by sending $a + J \longmapsto f(a)$. Then it is clearly well-defined (cf. the proof of Theorem 2.3), its kernel is I/J, which one can easily verify, and the commutativity of the diagram is automatic. Moreover, the defined map making the diagram commutative is automatically unique because π is surjective. The proof of Part (2) is too easy. □

The following theorem describes the fundamental relations between ideals in a quotient ring R/J and those ideals I of R containing the ideal J.

THEOREM 5.4. *(Correspondence Theorem) Let* J *be an ideal of a ring* R *and* $\pi : R \longrightarrow R/J$ *be the associated canonical map. Then*

(1) *There exists a one-to-one correspondence between the set of those ideals* I *containing* J *and the set of all ideals of* R/J*, given by*

$$I \longmapsto \pi(I) = I/J \qquad \text{and} \qquad \bar{I} := I/J \longmapsto \pi^{-1}(\bar{I}) = I.$$

(2) *If* $I \subset R$ *corresponds to* $\bar{I} := I/J$ *under this correspondence, then the quotient rings* R/I *and* $(R/J)/(I/J)$ *are isomorphic:*

$$R/I \approx (R/J)/(I/J).$$

PROOF. Proof of part (1). We first note that both $\pi(I) = I/J$ and $\pi^{-1}(\bar{I}) = I$ hold (Exercise).To establish such a correspondence in Part (1), we need to verify the followings:

(i): If I is an ideal of R containing J, then its image $\pi(I) = I/J$ is an ideal of R/J.

(ii): If K' is an ideal of R/J, then its inverse image $\pi^{-1}(K')$ is an ideal of R containing J.

(iii): Every ideal K' of R/I is of the form I/J for some ideal I of R containing J.

(iv): $\pi^{-1}(\pi(I)) = I$ and $\pi(\pi^{-1}(\bar{I})) = \bar{I}$.

Proof of (i). We recall that the image of a subgroup is a subgroup (Lemma 2.18); hence we only need to prove that $\pi(I)$ is closed under scalar multiplication by elements of R/J but this is very easy: Let $\bar{r} \in R/J$ and $\bar{x} \in \pi(I)$. Thus, $\bar{r} = \pi(r)$ and $\bar{x} = \pi(x)$ for some $r \in R$ and $x \in I$. Then since π is a ring homomorphism $\bar{r}\bar{x} = \pi(r)\pi(x) = \pi(rx)$ with $rx \in I$, and consequently $\bar{r}\bar{x} \in \pi(I)$.

Proof of (ii) and proof of part (2). We consider the composition,

$$\varphi \circ \pi : R \xrightarrow{\pi} R/J \xrightarrow{\varphi} (R/J)/K'.$$

Since both π and φ are surjective, so is their composition $\varphi \circ \pi$; moreover, its kernel consists of those elements $r \in R$ such that $\pi(r) \in K' = \ker(\varphi)$. Thus,

$$\ker(\varphi \circ \pi) = \pi^{-1}(K'),$$

and hence $\pi^{-1}(K')$ is an ideal of R, and it clearly contains $\ker(\pi) = J$. This proves (ii). In particular, by the First Isomorphism Theorem applied to the composition $\varphi \circ \pi$, we see that $R/\pi^{-1}(\bar{I})$ is isomorphic to $(R/J)/\bar{I}$.

Proof of (iii). By (ii), $\pi^{-1}(K')$ is an ideal of R, say, I. Then $\pi^{-1}(K')$ clearly contains $\ker(\pi) = J$, and we have

$$K' = \pi(\pi^{-1}(K')) = \pi(I) = I/J,$$

because π is surjective.

Proof of (iv). We recall that the two inclusions

$$I \subseteq \pi^{-1}(\pi(I)) \qquad \text{and} \qquad \pi(\pi^{-1}(\bar{I})) \subseteq \bar{I}$$

are general properties of any map from a set to a set. We also recall that since π is surjective,

$$\pi(\pi^{-1}(\bar{I})) = \bar{I}$$

generally holds. Thus, to finish the proof, we need to prove that $\pi^{-1}(\pi(I)) \subseteq I$ and this requires the hypothesis that $J \subseteq I$. For this, let $x \in \pi^{-1}(\pi(I))$. Then $\pi(x) \in \pi(I)$; hence there exists an element $y \in I$ such that $\pi(x) = \pi(y)$. Thus, $0 = \pi(x) - \pi(y) = \pi(x - y)$; hence $x - y \in \ker \pi = J$. Then since $J \subseteq I$ and $y \in I$, we see that $x - y \in I$ and hence $x \in I$. This finishes the whole proof of the theorem. □

Exercises for Chapter 5.5

EXERCISE 5.33. *Complete the proof of Theorem 5.3.*

EXERCISE 5.34. *Let I, J be ideals of a ring R. Prove that if $I + J = R$ then $IJ = I \cap J$.*

EXERCISE 5.35. *Let I, J be ideals of a ring R. Prove that some power of the image of each element of $I \cap J$ in R/IJ is zero.*

EXERCISE 5.36. *(Chinese Remainder Theorem) Let I, J be ideals of a ring R. Prove that for any pair a, b of elements of R there exists an element z such that $z - a \in I$ and $z - b \in J$.*

6. Relations Among Elements in a Ring

The construction of quotient rings in the previous Section has an important interpretation in terms of relations among elements in a ring R. The following procedure of introducing relations among elements of a ring R will lead to a new ring in most cases, and this makes such a procedure very important.

Suppose that we perform a sequence of operations $+, -, \times$ on some elements of R to get a new element, say, a.

DEFINITION 5.34. *If the resulting element a is zero, we say that the given elements are* ***related in R by the equation***

$$a = 0. \tag{5.8}$$

EXAMPLE 5.25. *The numbers $2, 3$, and 6 are related in $\mathbb{Z}$ by the equation*

$$2 \times 3 - 6 = 0.$$

Now if the new element a obtained by a sequence of operations $+, -, \times$ on some elements of R is not zero, we may ask whether it is possible to modify the ring R itself in such a way that $a = 0$ in the modification of R.

EXAMPLE 5.26. *The relation* $3 \times 4 - 5 = 0$ *does not hold in* $\mathbb{Z}$*, because the left-hand side is* 7*. But if we impose the relation* $7 = 0$ *on the integers, then the left-hand side becomes* 0*, and doing so amounts to working modulo* 7*.*

Now let a be an arbitrary element of R, and we want to modify R in order to impose the relation $a = 0$, but we still want to keep the operations $+$ and $\times$ so that we will have to accept some consequences of this relation to happen, simply because we are doing Algebra. For instance, if we want to set $a = 0$, then we must also accept to happen: $ra = 0$ for any natural number r, and $b + a = b$, and hence $b + ra = b$.

We then note that Theorem 5.3 tells us that this is enough: There are no other consequences of (5.8). In order to see this, we first note that if we fix an element b but let r vary, then the set $\{b + ra\}$ is the coset

$$b + (a),$$

where $(a) = aR$ is the principal ideal generated by a. Thus, *setting* $b + ra = b$ *for all* r *is the same as equating the elements of this coset, and this is precisely what happens when we pass from* R *to the quotient ring*

$$\bar{R} = R/(a).$$

We recall that elements of $\bar{R}$ are the cosets $\bar{b} = b + (a)$, and the canonical map $\pi : R \longrightarrow \bar{R}$ carries all the elements $b + ra$ in one coset to the same element $\bar{b} = \pi(b)$ of $\bar{R}$. Thus, exactly the right amount of collapsing has taken place in $\bar{R}$. Moreover, $\bar{a} = 0$ holds in $\bar{R}$ because a is an element of the kernel, (a), of π. Therefore, *it is reasonable to view* $\bar{R} = R/(a)$ *as the ring obtained by introducing the relation* $a = 0$ *in* R*.*

Moreover, this reasoning does make sense because of the following explanation: If our element a was obtained from some other elements by a sequence of ring operations, as we supposed in (5.8), then the fact that π is a ring homomorphism guarantees the same sequence of operations gives 0 in $\bar{R}$. For instance, if $uv + w = a$ for some $u, v, w \in R$, then the relation

$$\bar{u}\bar{v} + \bar{w} = 0$$

holds in $\bar{R}$. For since π is a ring homomorphism, we have

$$\bar{u}\bar{v} + \bar{w} = \overline{uv + w} = \bar{a} = 0.$$

EXAMPLE 5.27. *A good example of this construction is the relation $n = 0$ in the ring $\mathbb{Z}$ of integers. The resulting ring is the quotient ring $\mathbb{Z}/n\mathbb{Z}$.*

More generally, we can introduce any number of relations $a_1 = 0, ..., a_n = 0$, by taking the ideal

$$I = (a_1, ..., a_n)$$

generated by $a_1, ..., a_n$. (Cf. (5.7)). Thus, by definition, I is the set of all possible linear combinations

$$I = \{r_1 a_1 + ... + r_n a_n : r_i \in R\}.$$

Then, as we have seen above, the quotient ring $\bar{R} = R/I$ should be viewed as the ring obtained by introducing the n relations $a_1 = 0, ..., a_n = 0$ into R. Since $a_i \in I$, the residues $\bar{a}_i$ in $\bar{R}$ are zero, and two elements b, b' of R have the same image in $\bar{R}$ if and only if $b' - b \in I$, or $b' = b + r_1 a_1 + ... + r_n a_n$, for some $r_i \in R$. Thus, the relations

$$b + r_1 a_1 + ... + r_n a_n = b,$$

for various r_i are the only possible consequences of $a_1 = 0, ..., a_n = 0$.

Moreover, it follows from the Third Isomorphism Theorem (part (2) of Theorem 5.4) that *introducing relations one at a time or all together leads to isomorphic results.* For instance, let a, b be two elements of a ring R. Then $\bar{R} = R/(a)$ is the result of killing a, and introducing the relation $\bar{b} = 0$ into the ring $\bar{R}$ leads to the quotient ring $\bar{R}/(\bar{b})$, and then by part (2) of Theorem 5.4, this quotient ring is isomorphic to the quotient ring

$R/(a, b)$ obtained by killing both a and b at the same time, because the ideal (a, b) in R and the ideal $(\bar{b})$ in $\bar{R}$ are corresponding ideals by part (1) of Theorem 5.4. More precisely, we note that $\alpha \in (\bar{b})$ if and only if $\alpha = rb + (a)$ for some $r \in R$ if and only if $\alpha = rb + sa + (a)$ for some $r, s \in R$ if and only if $\alpha \in (a, b)/(a)$. Thus, $(\bar{b}) = (a, b)/(a)$; hence $\bar{R}/(\bar{b})$ is isomorphic to $R/(a, b)$ by part (2) of Theorem 5.4.

Thus, the more relations we add, the more collapsing takes place in the map $R \longrightarrow \bar{R}$. But, if we add them carelessly, then the worst that can happen is that we may end up with $I = R$. For instance, all relations $a = 0$ become true when we collapse the ring R to the zero ring $\{0\}$.

In some simple cases the First Isomorphism Theorem (part (2) of Theorem 5.3) can be applied to relate the ring obtained to a more familiar one. In what follows, we will present two examples to demonstrate this.

Let $R = \mathbb{Z}[i]$ be the ring of Gauss integers, and let $\bar{R}$ be the ring obtained by introducing the relation,

$$1 + 3i = 0,$$

in R. Thus,

$$\bar{R} = R/(1 + 3i),$$

where $(1 + 3i)$ is the principal ideal of R generated by $1 + 3i$. As we have done above, we try to figure out algebraic consequences of this relation in R by doing a few algebraic manipulations on $-1 = 3i$. First of all, multiplying both sides of this equality by i, we see that

$$i = 3.$$

This means that $i = 3$ in $\bar{R}$. On the other hand, $i^2 = -1$ in R, and hence in $\bar{R}$ too. Therefore, $3^2 = -1$, or $10 = 0$ in $\bar{R}$. Then since $i = 3$ and $10 = 0$ in $\bar{R}$, it is quite reasonable to guess that $\bar{R}$ is isomorphic to $\mathbb{Z}/(10) = \mathbb{Z}/10\mathbb{Z}$. Then the following proposition actually proves that it is the case:

PROPOSITION 5.17. *The ring* $\mathbb{Z}[i]/(1+3i)$ *is isomorphic to the ring* $\mathbb{Z}/10\mathbb{Z}$ *of integers modulo* 10.

PROOF. By Proposition 5.12, we know that there exists a unique ring homomorphism

$$\varphi : \mathbb{Z} \longrightarrow \bar{R} = \mathbb{Z}[i]/(1+3i).$$

Then by the First Isomorphism Theorem (part (2) of Theorem 5.3), we know that $im(\varphi)$ is isomorphic to $\mathbb{Z}/\ker(\varphi)$. Thus, if we can prove that φ is surjective and that $\ker(\varphi) = 10\mathbb{Z}$, then we will have succeeded.

For this, we first recall that every element of $\bar{R}$ is the residue of a Gauss integer $a+bi$. Then since $i = 3$ in $\bar{R}$, the residue of $a+bi$ is the same as that of the integer $a+3b$; hence φ is surjective. Next, let n be an element of $\ker(\varphi)$. From the fact that $\bar{R} = R/I$, we easily see that n must be a multiple of $1+3i$ in the ring of Gauss integers. Thus, we may write

$$n = (a+bi)(1+3i) = (a-3b) + (3a+b)i$$

for some $a, b \in \mathbb{Z}$. Then since n is an integer, we must have

$$3a + b = 0,$$

or $b = -3a$. Thus,

$$n = a(1-3i)(1+3i) = 10a,$$

and this shows that

$$\ker(\varphi) \subseteq 10\mathbb{Z}.$$

On the other hand, we have already seen that

$$10 = (1-3i)(1+3i) \in \mathbb{Z}[i],$$

and hence $10 \in \ker(\varphi)$. Therefore, $\ker(\varphi) = 10\mathbb{Z}$, and this is what we wanted to prove. □

Thus, we easily see that

COROLLARY 5.14.

$$\mathbb{Z}[i]/(1+3i) \simeq \mathbb{Z}/10\mathbb{Z}\ (= \mathbb{Z}_{10}).$$

is not an integral domain.

Another possible way to identify the given quotient ring R/I with another ring is to find a ring R' and a ring homomorphism $\varphi : R \longrightarrow R'$ whose kernel is I. To demonstrate this, we consider $\bar{R} = \mathbb{C}[x, y]/(xy)$. Then we note that the principal ideal (y) in $\mathbb{C}[x, y]$ is the kernel of the substitution homomorphism $\mathbb{C}[x, y] \longrightarrow \mathbb{C}[x]$ by sending $y \longmapsto 0$. (Proposition 5.10). Since $\mathbb{C}[x]$ is a subring of $\mathbb{C}[x, y]$, the substitution homomorphism is surjective. Thus, we have a isomorphism

$$\mathbb{C}[x, y]/(y) \approx \mathbb{C}[x].$$

Similarly, we have an isomorphism: $\mathbb{C}[x, y]/(x) \approx \mathbb{C}[y]$. These results strongly suggest that there should exist a ring homomorphism

$$\varphi : \mathbb{C}[x, y] \longrightarrow \mathbb{C}[x] \times \mathbb{C}[y]$$

defined by sending $f(x, y) \longmapsto (f(x, 0), f(0, y))$.

PROPOSITION 5.18. *The quotient ring* $\mathbb{C}[x, y]/(xy)$ *is isomorphic to the subring of the product ring* $\mathbb{C}[x]\times\mathbb{C}[y]$ *consisting of those pairs* $(p(x), q(y))$ *such that* $p(0) = q(0)$.

PROOF. One can easily check that φ is a ring homomorphism with kernel,

$$\ker(\varphi) = (x) \cap (y) = (xy).$$

Thus by the First Isomorphism Theorem (part (2) of Theorem 5.3), $\bar{R} = \mathbb{C}[x, y]/(xy)$ is isomorphic to the image of the homomorphism φ. Then by definition of φ, any element

$$(p(x), q(y)) = (f(x, 0), f(0, y))$$

of $im(\varphi)$ necessarily satisfies the condition $p(0) = q(0)$, and moreover, given any pair $(p(x), q(y)) \in \mathbb{C}[x] \times \mathbb{C}[y]$ such that $p(0) = q(0)$, we see that $p(x)+q(y)-p(0) \in \mathbb{C}[x, y]$ is obviously mapped to $(p(x), q(y))$; hence $im(\varphi)$ consists of those pairs $(p(x), p(y))$ such that $p(0) = q(0)$, and this finishes the proof. □

We note that there are no general methods for identifying a quotient ring with another ring, except the First Isomorphism Theorem, because it will usually not be a familiar ring:

EXAMPLE 5.28. *The quotient ring*

$$\mathbb{C}[x, y]/(y^2 - x^3 + x)$$

is fundamentally different from any ring we have seen up to now. In fact, this ring (the coordinate ring of the elliptic curve $y^2 = x^3 - x$) is not a UFD (Unique Factorization Domain).

Exercises for Chapter 5.6

EXERCISE 5.37. *Prove that the quotient ring $\mathbb{Z}_{96}/6\mathbb{Z}_{96}$ is isomorphic to $\mathbb{Z}_6$.*

EXERCISE 5.38. *Prove that the quotient ring $\mathbb{Z}_{48}/4\mathbb{Z}_{48}$ is isomorphic to $\mathbb{Z}_4$.*

EXERCISE 5.39. *Prove that the quotient ring $\mathbb{Z}[i]/(i+2)$ is isomorphic to $\mathbb{Z}_5$.*

EXERCISE 5.40. *Let a be an element of a ring R and let R' be the ring obtained from R by adjoining an inverse of a. Prove that R' is the zero ring if and only if a is nilpotent.*

7. Adjunction of Elements

In this section we discuss a procedure of adding new elements to a ring, which we will soon see that it is closely related to the introduction of relations among elements in a ring R. Our model for this procedure is the construction of the complex field $\mathbb{C}$, starting from the real numbers $\mathbb{R}$: One obtains $\mathbb{C}$ from $\mathbb{R}$ by adding $i = \sqrt{-1}$, and the construction is completely formal. That is, the imaginary number $i = \sqrt{-1}$ satisfies no other properties but those forced by the relation

$$i^2 = -1. \tag{5.9}$$

We are now ready to understand the general principle behind this construction. Let R be an arbitrary ring, and a bigger ring R' which contains R as a subring and also contains a new element, say α. We will probably want α to satisfy some relations such as (5.9), for instance.

DEFINITION 5.35. *A ring R' containing R as a subring is called a* ***ring extension*** *of R.*

Thus, we are looking for a suitable ring extension of R.

For this, we start an algebraic relation, like the one for i in (5.9), for a not-yet determined new element α. Then, we basically have two different cases: If the element α happens to be an element of a ring extension R' of R, which we already know, the our solution is the subring of R' generated by R and α. This subring is denoted by

$$R[\alpha],$$

and we have already described this ring in Section One of this Chapter, in the case of $R = \mathbb{Z}$ and $R' = \mathbb{C}$. The description for elements of $R[\alpha]$ is no different in general: $R[\alpha]$ consists of those elements of R' which can be expressed as a polynomial in α:

$$r_n\alpha^n + \cdots + r_1\alpha + r_0$$

with coefficients $r_i \in R$.

Otherwise, that is, if the element α is not an element of a certain well-known ring extension R' of R, then we really have to construct a new ring R' such that R' contains the element α satisfying the imposed relation for α in R'.

REMARK 5.29. *The following method of constructing such a new ring R' is due to Kronecker (1823 - 1891).*

We first note that the polynomial ring $R[x]$ in one unknown x is a ring extension of R and it is generated by R and the unknown symbol x. Thus, the notation $R[x]$ agrees with the one introduced above. Moreover, the Substitution Principle (Proposition 5.10) tells us that the polynomial ring is the *universal solution* to our problem of adjoining a new element, in the following sense: If α is an element of a *not-yet determined* ring extension R' of R, then there exists a unique ring homomorphism $R[x] \longrightarrow R'$ which is identity on R and which carries the unknown symbol x to α. Then the image of this map will be the subring $R[\alpha]$ of R'.

Thus, we are left to abstractly construct the ring extension R' containing α. For this, let's consider the question of the relations we want our new element α to satisfy. The variable x in the polynomial ring $R[x]$ satisfies no relations except those, such as $0x = 0$, implied by the ring axiom, and this is another way to state the universal property of the polynomial ring. Then using the procedure described in the previous section with title, Relations among Elements in a Ring, we can add relations to the polynomial ring $R[x]$: *We introduce any relations among polynomials in $R[x]$ as we like by using the construction of quotient rings of the polynomial ring $R[x]$.*

EXAMPLE 5.29. *We can construct the complex numbers $\mathbb{C}$ formally by introducing the relation $x^2 + 1 = 0$ into the ring $\mathbb{R}[x]$ of real polynomials. For this, we form the quotient ring $\mathbb{R}[x]/(x^2 + 1)$. Then the residue of x under the canonical ring homomorphism $\pi : \mathbb{R}[x] \longrightarrow \mathbb{R}[x]/(x^2 + 1)$ becomes our element i, i.e., the one of which square is equal to -1. We note that the relation*

$$\overline{x^2 + 1} = \bar{x}^2 + \bar{1} = 0$$

holds in $\mathbb{R}[x]/(x^2 + 1)$, because the kernel of the canonical ring homomorphism $\pi : \mathbb{R}[x] \longrightarrow \mathbb{R}[x]/(x^2 + 1)$ is $x^2 + 1$. And since $\bar{1}$ is the identity element, i.e., the multiplicative identity, in $\mathbb{R}[x]/(x^2 + 1)$, we drop the usual notation for the multiplicative identity drops the bar. Thus, the quotient ring $\mathbb{R}[x]/(x^2 + 1)$ is obtained from $\mathbb{R}$ by adjoining an element $\bar{x}$ satisfying

$$(\bar{x})^2 + 1 = 0.$$

Moreover, one can easily check that $\mathbb{R}[x]/(x^2+1)$ *is a real vector space with basis* $\{1, i = \bar{x}\}$, *so that*

$$\mathbb{R}[x]/(x^2+1) = \{a + bi : a, b \in \mathbb{R}\}.$$

Furthermore, we note that for any non-zero $a + bi \in \mathbb{R}[x]/(x^2+1)$, *we have*

$$(a+bi)\left(\frac{a-bi}{a^2+b^2}\right) = 1,$$

so that any non-zero $a + bi$ *is invertible in* $\mathbb{R}[x]/(x^2+1)$. *Therefore, we have formally constructed the field* $\mathbb{C}$ *of complex numbers:* $\mathbb{R}[x]/(x^2+1) = \mathbb{C}$.

On the other hand, suppose that the field

$$\mathbb{C} := \{a + bi : a, b \in \mathbb{R}\}$$

of complex numbers is constructed by another means. Then we can prove that there exists an isomorphism $\mathbb{R}[x]/(x^2+1) \approx \mathbb{C}$ *using the First Isomorphism Theorem (part (2) of Theorem 5.3): Substitution (Proposition 5.10) of* i *for* x *defines a surjective ring homomorphism*

$$\varphi : \mathbb{R}[x] \longrightarrow \mathbb{C},$$

whose kernel is the set of those real polynomials with i *as a root. We note that if* i *is a root of a real polynomial* $p(x)$, *then* $-i$ *also its root; hence* $x - i$ *and* $x + i$ *both divide* $p(x)$. *Thus, the kernel consists of those polynomials divisible by* $(x-i)(x+i) = x^2+1$, *that is, the kernel is the principal ideal* (x^2+1). *Then by the First Isomorphism Theorem (part (2) of Theorem 5.3), we conclude that* $\mathbb{R}[x]/(x^2+1) \approx \mathbb{C}$.

Here is another example:

EXAMPLE 5.30. *We say that an element* ϵ *of a ring* R *is* ***infinitesimal*** *or* ***nilpotent*** *if some power* $\epsilon^n = 0$. *We want to introduce the relation,*

$$\epsilon^2 = 0$$

in a ring R. Thus, the result of adjoining an element ϵ such that $\epsilon^2 = 0$ to a ring R is the quotient ring

$$R[x]/(x^2),$$

where the residue of x is the infinitesimal element ϵ. We note that this ring is not a field because ϵ has no inverse in it. It is even not an integral domain. Moreover, in this quotient ring, the relation $\epsilon^2 = 0$ reduces all polynomial expressions in ϵ to degree < 2, so that the elements of $R[x]/(x^2)$ are of the form,

$$a + b\epsilon,$$

with $a, b \in R$. Here multiplication and addition of two elements of $R[x]/(x^2)$ are given by the rules:

$$\begin{aligned}(a + b\epsilon)(c + d\epsilon) &= ac + (bc + ad)\epsilon, \\ (a + b\epsilon) + (c + d\epsilon) &= (a + c) + (b + d)\epsilon,\end{aligned}$$

where each of them are naturally induced from those two operations for polynomials in $R[x]$ modulo $x^2 = 0$.

In general, if we want to adjoin an element α satisfying one or more polynomial relations of the form

$$f(\alpha) = c_n\alpha^n + \cdots + c_1\alpha + c_0 \tag{5.10}$$

to a ring R, the solution is the quotient ring $R[x]/I$, where I is the ideal in $R[x]$ generated by the polynomials $f(x)$. If α denotes the residue $\bar{x}$ of x in $R[x]/I$, then we have

$$0 = \overline{f(x)} = \bar{c}_n\bar{x}^n + \cdots + \bar{c}_0 = \bar{c}_n\alpha^n + \cdots + \bar{c}_0,$$

where each $\bar{c}_i$ is the image in $R[x]/I$ of the constant polynomial c_i. Thus α satisfies the relation in $R[x]/I$ which corresponds to the relation (5.10) in R.

DEFINITION 5.36. *The ring $R[x]/I$ obtained in this way is usually denoted by*

$$R[\alpha] = \textit{ring obtained by } \textbf{\textit{adjoining}} \ \alpha \textit{ to } R.$$

REMARK 5.30. *If R is a field, then R is a subring of $R[x]/I$ by Corollary 5.11. If R is not a field, then R is not a subring of $R[x]/I$ in general, unless I is the principal ideal $(f(x))$ generated by a* monic *polynomial $f(x) \in R[x]$. (Proposition 5.19).*

Several $\alpha_1, ..., \alpha_m$ can be adjoined repeating this procedure, or by introducing the appropriate relations in the polynomial ring $R[x_1, ..., x_m]$ in m unknowns all at once.

One of the most important cases is that the new element α is required to satisfy a single *monic* equation of degree $n > 0$. Suppose that we want the relation $f(x) = 0$, where $f(x)$ is the monic polynomial

$$f(x) = x^n + c_{n-1}x^{n-1} + \cdots + c_1 x + c_0.$$

LEMMA 5.19. *Let $f(x)$ be a* monic *polynomial of degree n. Then principal ideal $(f(x))$ contains no non-zero polynomial of degree $< n$.*

PROOF. By definition the ideal $(f(x))$ consists of all multiples $h(x)f(x)$ of $f(x)$. We write

$$h(x) = b_m x^m + \cdots + b_1 x + b_0, \qquad b_m \neq 0$$

and then the leading term, i.e., the highest-degree term, of $h(x)f(x)$ is $b_m x^{m+n}$, and hence $\deg(h(x)f(x)) = m + n \geq n$. □

PROPOSITION 5.19. *Let R be a ring, and let $f(x)$ be a* monic *polynomial of positive degree n with coefficients in R. Then elements of the ring $R[\alpha]$ obtained by adjoining an element α satisfying the relation $f(\alpha) = 0$ are in one-to-one correspondence with vectors $(r_0, ..., r_{n-1}) \in R^n$ by the assignment:*

$$(r_0, ..., r_{n-1}) \longleftrightarrow r_0 + r_1\alpha + r_2\alpha^2 + \cdots + r_{n-1}\alpha^{n-1},$$

where $r_i \in R$. *In particular,* R *is a subring of* $R[\alpha] = R[x]/(f(x))$.

PROOF. We recall that every polynomial in $R[x]$ is uniquely expressed as a linear combination of monomials x^i with coefficients in R, that is, the monomials x^i is a basis of $R[x]$ over R. Thus their images α^i modulo $f(\alpha)$ do the same role for

$$R[\alpha] = R[x]/(f(x)),$$

i.e., $1, \alpha, \alpha^2, ..., \alpha^{n-1}$ form a generating set for $R[\alpha]$ over R. More precisely, we know that every element of $R[\alpha] = R[x]/(f(x))$ is the residue of a polynomial in $R[x]$; hence it can be written in the form $g(\alpha)$ for some polynomial $g(x) \in R[x]$. Then the relation

$$f(\alpha) = 0$$

can be used to replace any polynomial $g(\alpha)$ of degree $\geq n = \deg(f(x))$ by one of lower degree. In order to explicitly figure out $g(\alpha)$, we apply division with remainder by $f(x)$ on the polynomial $g(x)$, which results in

$$g(x) = q(x)f(x) + r(x)$$

with $0 \leq \deg(r(x)) < \deg(f(x))$ by Proposition 5.15. Then since $f(\alpha) = 0$, $g(\alpha) = r(\alpha)$, and hence every element β of $R[\alpha]$ is written as a polynomial in α, of degree $< n$.

Now, if there were another expression for the same element β of $R[\alpha]$,

$$r'_0 + r'_1\alpha + \cdots + r'_{n-1}\alpha^{n-1},$$

then their difference is zero so that the polynomial

$$d(x) = (r_0 - r'_0) + (r_1 - r'_1)x + (r_2 - r'_2)x^2 + \cdots + (r_{n-1} - r'_{n-1})x^{n-1}$$

in $R[x]$ belongs to the principal ideal $(f(x))$ of $R[x]$, and hence by Lemma 5.19 $d(x)$ must be the zero polynomial, i.e., each $r_i - r'_i = 0$ for $i = 1, 2, ..., n-1$, because

$$\deg(f(x)) = n > n - 1 = \deg(d(x)).$$

Thus, $1, \alpha, \alpha^2, ..., \alpha^{n-1}$ are linearly independent over R. This establishes the claimed one-to-one correspondence.

Now, the kernel of the composition of the two ring homomorphisms

$$R \hookrightarrow R[x] \longrightarrow R[x]/(f(x))$$

is $(f(x)) \cap R$, so that it consists of constant polynomials c in the ideal $(f(x))$. Thus, $c(\alpha) = c = 0$ in $R[x]/(f(x))$ only when $c = 0$, and this shows that R is a subring of $R[x]/(f(x))$. □

REMARK 5.31. *We note that addition of two such expressions*

$$r_0 + r_1\alpha + r_2\alpha^2 + \cdots + r_{n-1}\alpha^{n-1}$$

in $R[\alpha]$ is the usual vector addition so that it only depends on the degree, but multiplication of two such expressions is a little bit complicated: Multiplication of two such expressions $r_0+r_1\alpha+r_2\alpha^2+\cdots+r_{n-1}\alpha^{n-1}$ in $R[\alpha]$ is induced from those for polynomials in $R[x]$ via the modulo computation by $f(x)$, that is, to multiply two such linear combinations in $R[\alpha]$ we use polynomial multiplication and then divide the product by $f(x)$; then the remainder is the linear combination of $1, \alpha, \alpha^2, ..., \alpha^{n-1}$ which represents the product in $R[\alpha]$. Thus, multiplication is strongly dependent on the particular polynomial $f(x)$.

We now present a few examples of adjoining a new element to a ring R:

EXAMPLE 5.31. *Let $\mathbb{Z}[\alpha]$ be the result of adjoining an element α to a ring $\mathbb{Z}$ satisfying the relation,*

$$\alpha^3 + 3\alpha + 1 = 0.$$

Thus,

$$\mathbb{Z}[\alpha] = \mathbb{Z}[x]/(x^3 + 3x + 1),$$

according to what we have studied in this section.

EXAMPLE 5.32. *Let* $\mathbb{Z}_5[\alpha]$ *be the result of adjoining an element* α *to a field*

$$\mathbb{Z}_5 = \mathbb{Z}/5\mathbb{Z} = \{\bar{0}, \bar{1}, \bar{2}, \bar{3}, \bar{4}\},$$

satisfying the relation

$$\alpha^2 - 3 = 0.$$

Thus,

$$\mathbb{Z}_5[\alpha] = \mathbb{Z}_5[x]/(x^2 - 3).$$

By Proposition 5.19, we see that $1, \alpha$ *are a basis of* $\mathbb{Z}_5[\alpha]$ *over the field* $\mathbb{Z}_5$ *and*

$$\mathbb{Z}_5[\alpha] = \{a + b\alpha : a, b \in \mathbb{Z}_5\}.$$

Thus, it consists of $5 \cdot 5 = 25$ *elements.*

In particular,

CLAIM 5.1. $\mathbb{Z}_5[\alpha]$ *is a field.*

For this, we first note that

$$(a + b\alpha)(a - b\alpha) = a^2 - 3b^2 \in \mathbb{Z}_5,$$

which is called the ***norm*** *of* $a + b\alpha$*. To prove the claim, we need to show that each non-zero* $a + b\alpha$ *is invertible in* $\mathbb{Z}_5[\alpha]$*, since it is already a commutative ring.*

Case 1. If $a \neq 0$ *and* $b = 0$*, then* a *is invertible in* $\mathbb{Z}_5$*, so that* $a + b \cdot 0$ *is invertible in* $\mathbb{Z}_5[\alpha]$*.*

Case 2. If $a = 0$ *and* $b \neq 0$ *then* $(b\alpha) \cdot (-b\alpha) = -3b^2$*, and hence* $(b\alpha) \cdot \left(\frac{\alpha}{3b}\right) = 1$*, i.e.,* $b\alpha$ *is invertible in* $\mathbb{Z}_5[\alpha]$*.*

Case 3. If $a \neq 0$ *and* $b \neq 0$*, then since the equation* $x^2 = 3$ *has no solution in* $\mathbb{Z}_5$*, we see that* $a^2 - 3b^2 \neq 0$ *in* $\mathbb{Z}_5$*, otherwise,* $3 = \left(\frac{a}{b}\right)^2$*, an absurd. This shows that if* $a \neq 0$ *and* $b \neq 0$ *then* $a^2 - 3b^2$ *is invertible in* $\mathbb{Z}_5$ *and hence it is invertible in* $\mathbb{Z}_5[\alpha]$*.*

In summary, we have shown that the inverse of a non-zero $a + b\alpha$ *in* $\mathbb{Z}_5[\alpha]$ *is:*

$$(a^2 - 3b^2)^{-1}(a - b\alpha). \qquad \square$$

In fact, since $x^2 - 3 = 0$ *has no solution in* $\mathbb{Z}_5$, $x^2 - 3$ *is irreducible over* $\mathbb{Z}_5$ *so that the principal ideal* $(x^2 - 3)$ *is a maximal ideal of the Principal Ideal Domain* $\mathbb{Z}_5[x]$, *and hence* $\mathbb{Z}_5[x]/(x^2 - 3)$ *is a field containing* $\mathbb{Z}_5$ *as its subfield. For more about this, the readers may go to the Section, titled with Prime Ideals and Maximal Ideals, below in this Chapter.*

EXAMPLE 5.33. *If we apply the same procedure to* $\mathbb{Z}_{11}$ *of adjoining an element* α *satisfying the relation* $\alpha^2 - 3 = 0$, *we obtain the quotient ring,*

$$\mathbb{Z}_{11}[x]/(x^2 - 3).$$

But this ring is not a field, because

$$x^2 - 3 = (x + 5)(x - 5)$$

in $\mathbb{Z}_{11}[x]$ *so that*

$$(\alpha + 5)(\alpha - 5) = 0$$

is true in $\mathbb{Z}_{11}[x]/(x^2 - 3)$. *Thus, this ring is even not an integral domain. On the other hand, the two solutions of* $x^2 - 3$ *are* $\alpha = \pm 5 \in \mathbb{Z}_{11}$, *and hence the quotient ring* $\mathbb{Z}_{11}[x]/(x^2 - 3)$ *is not the one we are looking for as* $\mathbb{Z}_{11}[\alpha]$. *In this case,* $\mathbb{Z}_{11}[\alpha] = \mathbb{Z}_{11}$.

The main reason why we have this problem is that the polynomial $x^2 - 3 = (x + 5)(x - 5)$ *over* $\mathbb{Z}_{11}$ *so it is not irreducible over* $\mathbb{Z}_{11}$. *This example thus tells us the following:*

REMARK 5.32. *The relation* $f(\alpha) = 0$ *for adjoining an element* α *to a ring* R *needs to be of least degree in* α. *That is, the corresponding polynomial* $f(x)$ *must be irreducible over the bottom ring* R.

It is in general difficult to analyze the structure of the ring obtained by adjoining an element which satisfies a non-monic polynomial relation. But, one of the simplest and most exceptional cases is the adjunction of a multiplicative inverse of an element of R to R: If an element $a \in R$ is going to be invertible, then it should satisfy the relation

$$a\alpha - 1 = 0.$$

Thus, we can adjoin an inverse of a by forming the quotient ring

$$R[\alpha] = R[x]/(ax - 1).$$

Then any element

$$\beta = r_0 + r_1\alpha + \cdots + r_n\alpha^n \in R[\alpha]$$

satisfies $a\alpha = 1$; consequently

$$\beta = \alpha^n(r_0a^n + r_1a^{n-1} + \cdots + r_n).$$

Thus, every element β of $R[\alpha] = R[x]/(ax - 1)$ has the form

$$\beta = \alpha^n r,$$

for some $r \in R$ and for some non-negative integer n. In particular, this also explains why this ring has no basis of the type described in Proposition 5.19,

EXAMPLE 5.34. *One important example of such a ring is when R is a polynomial ring $F[t]$ over a field F and we want to adjoin an inverse to the unknown t. The required ring is then*

$$F[t, x]/(xt - 1).$$

This ring is naturally identified with the ring

$$F[t, t^{-1}]$$

*of **Laurent polynomials** in t. A Laurent polynomial is a polynomial in t and t^{-1} of the form*

$$f(t) = \sum_{i=-n}^{n} a_i t^i = a_{-n}t^{-n} + a_{-n+1}t^{-n+1} + \cdots + a_0 + a_1 t + \cdots + a_n t^n.$$

EXERCISE 5.41. *Construct the isomorphism*

$$F[t, x]/(xt - 1) \approx F[t, t^{-1}].$$

We note that we'd better not adjoin an inverse of 0 to a ring R. This is because, the relation $0\alpha = 1$ in a ring implies $0 = 1$ in there by Lemma

5.6, and hence the zero element is invertible only in the zero ring $\{0\}$. Thus, if we insist on adjoining an inverse of 0 to a ring R, then we must end up with the zero ring.

More generally, let a, b be two non-zero elements of a ring R such that their product $ab = 0$. If a is invertible in R, then

$$b = a^{-1}ab = a^{-1}0 = 0.$$

Equivalently, if $ab = 0$ then a is not invertible unless $b = 0$. From this we see that *if a product ab of two non-zero elements of a ring R is zero, then the procedure of adjoining an inverse of a to R must kill b.* This is indeed the case because the residue of b in the ring $R[x]/(ax-1)$ is zero:

$$b = 0x + b = -bax + b = -b(ax-1) \in (ax-1).$$

In particular, R is not a subring of $R[x]/(ax-1)$ in this case. Here, we note that $ax-1$ is not a monic polynomial over R.

EXAMPLE 5.35. *We know that $\bar{2} \cdot \bar{3} = 0$ in the quotient ring*

$$\mathbb{Z}/6\mathbb{Z}.$$

If we adjoin $\bar{3}^{-1}$ to this ring, we must kill $\bar{2}$ by the above observation. This means that the elements $\bar{4} = \bar{2} + \bar{2}, \bar{3} = \bar{2} + \bar{1}$, and $\bar{5} = \bar{2} + \bar{2} + \bar{1}$ in $\mathbb{Z}/6\mathbb{Z} = \mathbb{Z}_6$ collapse to $0, \bar{1}$, and $\bar{1}$, respectively. That is, $\mathbb{Z}/6\mathbb{Z}$ collapses to the field $\mathbb{Z}/2\mathbb{Z} = \mathbb{Z}_2$. Thus, the quotient ring we are looking for must be

$$\mathbb{Z}_6[x]/(\bar{3}x - \bar{1}) \approx \mathbb{Z}_2.$$

Here is a detailed proof of this isomorphism. First of all, we note that the kernel of the surjective ring homomorphism:

$$\mathbb{Z}[x] \xrightarrow{\varphi} \mathbb{Z}_6[x] \xrightarrow{\phi} \mathbb{Z}_6[x]/(\bar{3}x - \bar{1})$$

is generated by $\ker(\varphi)$ *and the inverse image of* $\ker(\phi) = (\bar{3}x - \bar{1})$, *so it is equal to the ideal* $(6, 3x-1)$ *of* $\mathbb{Z}[x]$, *and this establishes an isomorphism,*

$$\mathbb{Z}_6[x]/(\bar{3}x - \bar{1}) \approx \mathbb{Z}[x]/(6, 3x-1).$$

Now from the two relations, $6 = 0$ *and* $3x - 1 = 0$, *we note that* $6x = 0, 6x - 2 = 0$, *and hence* $2 = 0$. *Thus* $2x = 0$ *too. When we combine this relation with* $3x - 1 = 0$, *we furthermore see that* $0 = 3x - 1 = 2x + x - 1 = x - 1$. *Thus,* 2 *and* $x - 1$ *are in the kernel,* $(6, 3x - 1)$, *of the ring homomorphism,*

$$\mathbb{Z}[x] \longrightarrow \mathbb{Z}[x]/(6, 3x - 1).$$

The point is that this exactly means the ideal $(6, 3x - 1)$ *of* $\mathbb{Z}[x]$ contains *the ideal* $(2, x - 1)$. *On the other hand,* 6 *and* $3x - 1 = 3(x - 1) + 2$ *are obviously elements of the ideal* $(2, x - 1)$. *This proves that the two ideals are the same:*

$$(6, 3x - 1) = (2, x - 1).$$

Therefore, we have the claimed isomorphism:

$$\mathbb{Z}[x]/(6, 3x - 1) = \mathbb{Z}[x]/(2, x - 1) \approx \mathbb{Z}_2[x]/(x - 1) \approx \mathbb{Z}_2,$$

where the last isomorphism is obvious.

DEFINITION 5.37. *A non-zero element* a *of a ring* R *is a* ***zero divisor*** *if there exists a non-zero element* b *such that* $ab = 0$.

EXAMPLE 5.36. *The residue* $\bar{3}$ *of* 3 *in the quotient ring* $\mathbb{Z}/6\mathbb{Z}$ *is a zero-divisor in there.*

Exercises for Chapter 5.7

EXERCISE 5.42. *Prove that the quotient ring* $\mathbb{Z}[x]/(3, x)$ *is isomorphic to* $\mathbb{Z}_3$.

EXERCISE 5.43. *Prove that the quotient ring* $\mathbb{Z}[x]/(6)$ *is isomorphic to* $\mathbb{Z}_6[x]$.

EXERCISE 5.44. *Let* $F[x]$ *be the polynomial over a field* F.

(1) *Prove that* $F[x]/(x^2 - 1)$ *is isomorphic to* $F[x]/(x^2 - 4)$.
(2) *Prove that* $F[x]/(x^2 + 1)$ *is isomorphic to* $F[x]/(x^2 + 2x + 2)$.

EXERCISE 5.45. *Determine the structure of the ring obtained by adjoining an element* α *to* $\mathbb{Z}$ *satisfying each of the following relation(s):*

(1) $2\alpha - 6 = 0$.
(2) $2\alpha - 5 = 0$.
(3) $2\alpha - 6 - 0$ *and* $\alpha - 10 = 0$.

EXERCISE 5.46. *Prove that* $1, x-\alpha, (x-\alpha)^2, ..., (x-\alpha)^{n-1}$ *is a* $\mathbb{C}$*-basis for the quotient ring* $\mathbb{C}[x]/((x-\alpha)^n)$.

EXERCISE 5.47. *Let* α *denote the image of* x *in the quotient ring*

$$\mathbb{Z}[x]/(x^4 + x^3 + x^2 + x + 1).$$

Express the product $(\alpha^3+\alpha^2+\alpha)\alpha^5$ *as a* $\mathbb{Z}$*-linear combination of the basis* $(1, \alpha, \alpha^2, \alpha^3, \alpha^4)$.

8. Integral Domains and Fraction Fields

In this section, we are going to discuss the problem of embedding a given ring R as a subring into a field F. In the last section, we have seen that we can not adjoin the inverse of a zero-divisor unless we kill some of its elements. Thus, a ring which contains zero-divisors can not be embedded into a field.

DEFINITION 5.38. *An* ***integral domain*** *(or simply,* ***domain****)* R *is a non-zero (commutative) ring with no zero-divisors.*

EXAMPLE 5.37. *Any subring of a field is an integral domain.*

LEMMA 5.20. *Let* R *be a commutative ring. Then the "cancellation law" holds in* R *if and only if* R *has no zero-divisors.*

PROOF. Suppose that R satisfies the cancellation law: If $ab = ac$ and $a \neq 0$, then $b = c$. If $ab = 0$ and $a \neq 0$, then since

$$ab = 0 = a \cdot 0$$

holds, we cancel out a from both sides to conclude that $b = 0$,Thus, there are no non-zero b satisfying $ab = 0$ where $a \neq 0$. Conversely, it has no zero-divisors, it exactly means that $ab = 0$ implies either $a = 0$ or $b = 0$. Suppose that $ab = ac$ and $a \neq 0$. Then we have

$$a \cdot (b - c) = 0.$$

Then since $a \neq 0$ by hypothesis, $b - c$ must be zero, i.e., $b = c$, which is all about the cancellation law. □

REMARK 5.33. *An integral domain R is very special in the sense: Even when R is not a field so that the inverse of a may not be an element of R we can* cancel out *a non-zero element a from the equality $ab = ac$ to conclude that $b = c$, as if a were invertible in R.*

PROPOSITION 5.20. *If R is an integral domain, then so is the polynomial ring $R[x]$.*

PROOF. Let $f(x) = a_n x^n + \cdots + a_0$ where $a_n \neq 0$ and $g(x) = b_m x^m + \cdots + b_0$ where $b_m \neq 0$ be two non-zero polynomials in $R[x]$. Then the product of the leading coefficient a_n of $f(x)$ and the leading coefficient b_m of $g(x)$ is non-zero, so that $f(x)g(x)$ is not the zero polynomial. □

PROPOSITION 5.21. *An integral domain R with finitely many elements is a field.*

PROOF. Let R consist of distinct n elements $b_1, ..., b_n$. Then for any non-zero element a of R, we consider the n products,

$$ab_1, ab_2, ..., ab_n.$$

If some two of them are the same, say $ab_i = ab_j$, then the cancellation law tells us that $b_i = b_j$, but this is impossible. Thus, they are all distinct and thus some ab_i must be equal to the multiplicative identity 1. Thus, R is a field. □

Let R be an integral domain. In what follows, we are going to construct the "fraction field" of R. Our model for this is the construction of the rational numbers $\mathbb{Q}$ as fractions of integers. Once the idea of using fractions is brought up explicitly, the construction follows that of the rational number very closely. In constructing the fraction field of R, this construction is simpler than the method of adjoining inverses of all non-zero elements of R, of which procedure has been fully explained in the last section.

DEFINITION 5.39. *Let any two elements $a, b \neq 0$ be two elements of an integral domain R.*

(1) *A* ***fraction*** *is a symbol denoted by*

$$a/b.$$

(2) *Two fractions a_1/b_1 and a_2/b_2 are called* ***equivalent*** *if*

$$a_1b_2 = a_2b_1$$

is true in R, and is denoted by

$$a_1/b_1 \sim a_2/b_2.$$

REMARK 5.34. *We may instead denote a fraction by an element (a, b) of the Cartesian product $R \times R$, and then $(a_1, b_1) \sim (a_2, b_2)$ if and only if $a_1b_2 = a_2b_1$ in R.*

LEMMA 5.21. *The equivalent relation defined above is an equivalence relation on the set of all fractions.*

PROOF. The reflexive property and the symmetric property are obvious. For the transitive property, suppose that $a_1/b_1 \sim a_2/b_2$ and $a_2/b_2 \sim a_3/b_3$ hold. Then,

$$a_1b_2 = a_2b_1 \qquad \text{and} \qquad a_2b_3 = a_3b_2.$$

Thus, we multiply b_3 to both sides of the first equality and b_1 to both sides of the second equality to get

$$a_1b_2b_3 = a_2b_1b_3 \qquad \text{and} \qquad a_2b_3b_1 = a_3b_2b_1.$$

Thus, $a_1b_2b_3 = a_3b_2b_1$. Then canceling out b_2 from both sides of this equality, we see that

$$a_1b_3 = a_3b_1.$$

That is, $a_1/b_1 \sim a_3/b_3$ holds. □

DEFINITION 5.40. *Let R be an integral domain.*

(1) *The* ***field of fractions*** *F of R is defined as the set of equivalence classes of fractions.*
(2) *Two fractions a_1/b_1 and a_2/b_2 are* **equal** *in F if they are equivalent fractions. That is,*

$$a_1/b_1 = a_2/b_2$$

in F if and only if $a_1b_2 = a_2b_1$ in R.
(3) *Addition and multiplication of two fractions are defined as follows:*

$$(a/b)(c/d) = (ac)/(bd), \qquad a/b + c/d = \frac{ad + bc}{bd}.$$

REMARK 5.35. *One can easily check that the two rules in (3) lead to equivalent answers if a/b and c/d are replaced by equivalent fractions.*

THEOREM 5.5. *Let R be an integral domain. Then the field F of fractions of R is a field, containing R as a subring, under addition and multiplication defined in the above Definition.*

PROOF. We leave it as an exercise to verify that F becomes a field under addition and multiplication defined in the above Definition. To show that R is a subring of F, we define a map: $R \longrightarrow F$ by sending

$$a \longmapsto a/1.$$

Then $a/1 = b/1$ if and only if $a = b$; hence this map is an injective ring homomorphism. Under this injection, R is identified as a subring of F consisting of those elements of the form $a/1$. □

EXAMPLE 5.38. *Let R be the polynomial ring $K[x]$, where K is a field. Since K is an integral domain, $K[x]$ is also an integral domain. Its field of fractions is called the* ***field of rational functions*** *in x, with coefficients in K. This field is usually denoted by*

$$K(x) = \left\{ \begin{array}{c} \text{equivalence classes of fractions } f(x)/g(x), \text{ where } f(x), g(x) \\ \text{are polynomials over } K \text{ and } g(x) \text{ is not the zero polynomial} \end{array} \right\}.$$

If $K = \mathbb{R}$, then *evaluation* of a rational function $f(x)/g(x)$ defines an actual function on the real line, whenever $g(x) \neq 0$. Again, as with polynomials, we should distinguish between the formally defined rational functions, which are fractions of polynomials, and the *actual* functions which they define by evaluation.

The fraction field is a universal solution to the problem of embedding an integral domain in to a field as is described in the following proposition:

PROPOSITION 5.22. *Let R be an integral domain, with field of fractions F, and let $\varphi : R \longrightarrow K$ be any* injective *ring homomorphism of R to a field K. Then there exists a* unique *ring homomorphism $\Phi : F \longrightarrow K$ whose restriction to R is φ, defined by sending*

$$a/b \longrightarrow \varphi(a)\varphi(b)^{-1}.$$

PROOF. Since φ is injective, $b \neq 0$ implies $\varphi(b) \neq 0$, so $\varphi(b)$ is invertible in K; hence $\varphi(a)\varphi(b)^{-1}$ is an element of K. Now, we need to verify that the map Φ is well-defined, namely, a different representation of the same equivalence class has the same image under Φ: Suppose that $a_1/b_1 \sim a_2/b_2$. Then, $a_1b_2 = a_2b_1$; hence $\varphi(a_1)\varphi(b_2) = \varphi(a_2)\varphi(b_1)$. Thus consequently

$$\Phi(a_1/b_1) = \varphi(a_1)\varphi(b_1)^{-1} = \varphi(a_2)\varphi(b_2)^{-1} = \Phi(a_2/b_2).$$

Then since φ is a ring homomorphism, Φ becomes automatically a ring homomorphism.

Moreover, any such ring homomorphism $\Psi : F \longrightarrow K$ whose restriction to R is φ must satisfy:

$$\begin{aligned}\Psi(a/b) &= \Psi(a/1 \cdot 1/b) = \Psi(a/1)\Psi(1/b) \\ &= \varphi(a)\varphi(b)^{-1} \\ &= \Phi(a/b).\end{aligned}$$

Thus it must be equal to Φ. □

Exercises for Chapter 5.8

EXERCISE 5.48. *Prove that a subring of an integral domain is an integral domain.*

EXERCISE 5.49. *Prove that a subring of a field is an integral domain.*

EXERCISE 5.50. *Prove that the polynomial ring* $R[x]$ *over an integral domain* R *is an integral domain.*

EXERCISE 5.51. *Prove that the invertible elements of the polynomial ring* $R[x]$ *over an integral domain* R *are the units in* R.

EXERCISE 5.52. *Prove that a subset of a finite field is a subfield if and only if it is closed under addition and multiplication and contains more than one element.*

EXERCISE 5.53. *Let* $d = 7$ *or* 11.

(1) *Prove that the set* $\mathbb{Q}(\sqrt{d})$ *of all real numbers* $a+b\sqrt{d}$ *for* $a, b \in \mathbb{Q}$ *is a field.*
(2) *Show that the function*

$$a + b\sqrt{7} \longmapsto a + b\sqrt{11}$$

is not a ring isomorphism of $\mathbb{Q}(\sqrt{7})$ *to* $\mathbb{Q}(\sqrt{11})$.
(3) *Prove that there is no ring isomorphism of* $\mathbb{Q}(\sqrt{7})$ *to* $\mathbb{Q}(\sqrt{11})$.

EXERCISE 5.54. *Prove that the fraction field of a field* F *is isomorphic to* F *itself.*

EXERCISE 5.55. *Prove that the fraction field of* $\mathbb{Z}[i]$ *is isomorphic to the ring of all complex numbers of the form,*

$$r + si,$$

where $r, s \in \mathbb{Q}$.

9. Prime Ideals and Maximal Ideals

We first recall

DEFINITION 5.41. *An ideal M of a ring R is called a* ***maximal ideal*** *if it is a proper ideal of R not contained in any other ideal, other than R and itself.*

PROPOSITION 5.23. *For any ideal I of a non-trivial ring R, the following statements are equivalent:*

(1) *If α, β are elements of R such that $\alpha\beta \in I$, then either $\alpha \in I$ or $\beta \in I$.*
(2) *If J, K are ideals of R such that $JK \subseteq I$, then either $J \subseteq I$ or $K \subseteq I$.*
(3) *The quotient ring R/I is an integral domain.*

PROOF. Part (2) clearly implies Part (1), if we take $J = (\alpha)$ and $K = (\beta)$.

The conditions for $\bar{R} = R/I$ to be an integral domain are that $\bar{R} \neq 0$ and $\bar{\alpha}\bar{\beta} = 0$ implies either $\bar{\alpha} = 0$ or $\bar{\beta} = 0$. We can translate these conditions back to: $I \neq R$ and if $\alpha\beta \in I$ then either $\alpha \in I$ or $\beta \in I$, and conversely. Thus, Part (1) and Part (3) are equivalent to each other.

Now, assume that Part (1) holds and let J, K be ideals of R such that $JK \subseteq I$. Suppose that J is not contained in I. Then there exists an element $\alpha \in J$ not contained in I. Let β be any element of K. Then $\alpha\beta \in JK$; hence $\alpha\beta \in I$. Thus, by hypothesis, $\beta \in I$. That is, we have shown that $K \subseteq I$. □

DEFINITION 5.42. *An ideal of R which satisfies one of these conditions is called a* ***prime ideal.***

PROPOSITION 5.24. *For any ideal I of a non-trivial ring R, we have*

(1) *I is a maximal ideal if and only if R/I is a field.*
(2) *I is a prime ideal if and only if R/I is an integral domain.*

PROOF. Part (1). By Proposition 5.13, we know that R is a field if and only if it has no proper ideals. Then since there is a natural one-to-one correspondence between the set of ideals of R containing I and the set of ideals of the quotient ring R/I (Theorem 5.4), we see that R/I is a field if and only if there are no ideal of R properly contained between R and I, namely, I is a maximal ideal of R.

The proof of Part (2) is immediate from the definition of an ideal I being prime: $ab \in I \Longrightarrow$ either $a \in I$ or $b \in I$, because this definition exactly says the cancellation law in the quotient ring: $(a+I)(b+I) = I \Longrightarrow$ either $a + I = I$ or $b + I = I$. □

PROPOSITION 5.25. *The maximal ideals of the ring $\mathbb{Z}$ of integers are the principal ideals generated by prime integers.*

PROOF. This is true because every ideal of $\mathbb{Z}$ is principal. □

Similarly, we know that the polynomial ring $k[x]$ in one variable over a field is a *PID*, by Proposition 5.16.

The maximal ideals of the ring $\mathbb{C}[x]$ of complex polynomials in one variable can also be described very simply:

Now, if the field k is moreover algebraically closed, for instance, $k = \mathbb{C}$, the field of complex numbers, we can say more:

PROPOSITION 5.26. (1) *The maximal ideals of the polynomial ring $\mathbb{C}[x]$ in one variable are principal ideals generated by the linear polynomials $x - \alpha$ in $\mathbb{C}[x]$.*

(2) *The ideal M_α generated by $x - \alpha$ is the kernel of the* ***evaluation homomorphism*** *ε_α:*

$$\begin{array}{c} \mathbb{C}[x] \\ \downarrow \varepsilon_\alpha \\ \mathbb{C} \end{array}$$

defined by

$$\varepsilon_\alpha(f(x)) = f(\alpha).$$

(3) *There is a one-to-one correspondence between maximal ideals*

$$M_\alpha = (x - \alpha)$$

of $\mathbb{C}[x]$ and complex numbers $\alpha \in \mathbb{C}$.

PROOF. Part (1). Let M be a maximal ideal in the *PID* $\mathbb{C}[x]$. Thus, it is generated by a single monic polynomial $f(x)$ of least degree over $\mathbb{C}$. Then since every complex polynomial of positive degree has a root α, $f(x)$ is divisible by the linear factor $x - \alpha$; hence $f(x)$ is contained in the principal ideal $(x - \alpha)$. Since M is an maximal ideal, we must have $M = (x - \alpha)$.

Part (2). The kernel, $\ker(\varepsilon_\alpha)$ is an ideal of $\mathbb{C}[x]$ obviously containing the linear polynomial $x - \alpha$, and other elements of $\ker(\varepsilon_\alpha)$ are clearly divisible by $x - a$, and hence $\ker(\varepsilon_\alpha) = (x - \alpha)$. We denote this maximal ideal by M_α. Part (3) then does not need a proof. □

REMARK 5.36. *Here is our comment on the role of the "evaluation homomorphism"* $\varepsilon_\alpha\colon \mathbb{C}[x] \longrightarrow \mathbb{C}$ *and the "evaluation function"* $f\colon \mathbb{C}^1 \longrightarrow \mathbb{C}$.

When we consider a function $g(x)$, *we usually assume that* x *is a variable for a fixed* g. *But, on the other hand, we can also think of a situation where* g *is a variable for a fixed* x, *whenever it makes sense mathematically.*

Thus, when $f(x)$ *is a given polynomial over* $\mathbb{C}$, *we can consider the evaluation function* $f\colon \mathbb{C}^1 \longrightarrow \mathbb{C}$, *which does make sense. And when* α *is a given point in* $\mathbb{C}^1$, *we can consider the evaluation homomorphism* $\varepsilon_\alpha\colon \mathbb{C}[x] \longrightarrow \mathbb{C}$, *which also makes sense. Therefore, in other words, the evaluation homomorphism* $\varepsilon_\alpha\colon \mathbb{C}[x] \longrightarrow \mathbb{C}$ *and the evaluation function* $f\colon \mathbb{C}^1 \longrightarrow \mathbb{C}$ *are* ***dual*** *to each other, related by the same value*

$$f(\alpha) = \varepsilon_\alpha(f) \in \mathbb{C}.$$

Exactly because of this reason, we sometimes define $\alpha(f)$ *to be* $\varepsilon_\alpha(f)$, *i.e.,*

$$\alpha(f) \equiv \varepsilon_\alpha(f) = f(\alpha),$$

so that we can easily move from one side of the duality to another side of the duality through this "local point evaluation process" in the bottom field $\mathbb{C}$.

In the above argument, we can safely replace $\mathbb{C}$ with any other ring R, without destroying the concept on the duality of the evaluation function and the evaluation homomorphism.

The extension of Proposition 5.26 to more than one variables is called the Null-stellen-satz (= Zero- Places-Theorem). Before we state and prove this important theorem, we need some lemmas and their consequences.

LEMMA 5.22. *Let V be a vector space which is spanned by a countable family $\{v_1, ..., v_n, ...\}$ of vectors. Then every set L of linearly independent vectors in V is either finite or countably infinite.*

PROOF. Let L be a linearly independent subset of V, V_n be the span of the first n vectors $v_1, ..., v_n$, and let

$$L_n = L \cap V_n.$$

Then L_n is a linearly independent subset in a finite-dimensional subspace V_n and hence it is a finite set. In addition to this, we easily see that

$$L = \cup_{n=1}^{\infty} L_n.$$

On the other hand, we recall that the union of countably many finite sets is either finite or countably infinite. □

EXAMPLE 5.39. *$k[x]$ is a k-vector space with a countable basis*

$$\{x^j : j \in \mathbb{N} \cup \{0\}\}.$$

Similarly, $k[x_1, ..., x_n]$ is a k-vector space with a countable basis

$$\{x_1^{\alpha_1} \cdots x_n^{\alpha_n} : \alpha_j \in \mathbb{N} \cup \{0\}\}.$$

We recall that the set $\mathbb{Q}$ of rational numbers is a countable infinite set, but the set $\mathbb{R}$ of real numbers are uncountable infinite set; hence so is the set $\mathbb{C}$ of complex numbers.

LEMMA 5.23. *The* uncountably *many rational functions*

$$\frac{1}{(x-\alpha)}, \qquad \alpha \in \mathbb{C},$$

are linearly independent over $\mathbb{C}$.

PROOF. We suppose that a linear combination of such rational functions:

$$\sum_{i=1}^{n} c_i \frac{1}{x-\alpha_i} = 0,$$

where $\alpha_1, ..., \alpha_n$ are distinct complex numbers with undetermined coefficients c_i. Then by clearing out the denominators, and by putting $x_i = c_i$ for each $i = 1, ..., n$, we easily see that each c_i must be zero. Thus, the rational functions $1/(x-\alpha)$ are linearly independent over $\mathbb{C}$. □

That is, we have equivalently shown that:

COROLLARY 5.15. *The field of fractions* $\mathbb{C}(x)$ *of the polynomial ring* $\mathbb{C}[x]$ *has a linearly independent subset consisting of uncountably many elements of* $\mathbb{C}(x)$. □

COROLLARY 5.16. *Let* M *be a maximal ideal of* $\mathbb{C}[x_1, ..., x_n]$, K *the field* $\mathbb{C}[x_1, ..., x_n]/M$, *and let* π_1 *be the restriction of the canonical projection*

$$\pi : \mathbb{C}[x_1, ..., x_n] \longrightarrow K$$

to the subring $\mathbb{C}[x_1]$ *of polynomials in one variable:*

$$\pi_1 \colon \mathbb{C}[x_1] \overset{i}{\hookrightarrow} \mathbb{C}[x_1, ..., x_n) \overset{\pi}{\longrightarrow} K, \qquad \pi_1 = \pi \circ i.$$

Then the kernel $\ker(\pi_1)$ *of* π_1 *is either* 0 *or else it is a maximal ideal of* $\mathbb{C}[x_1]$.

PROOF. Suppose that its kernel, $\ker(\pi_1)$, is not zero, and let $f(x_1)$ be a non-zero polynomial in $\ker(\pi_1)$. Since K is not the zero ring, $\ker(\pi_1)$ is not the whole ring. Then since we have the natural injective ring homomorphism $\mathbb{C} \longrightarrow K$, and any non-zero constant polynomial in $\mathbb{C}[x_1]$ remains the same under π_1; hence $f(x_1)$ is not a constant polynomial. Thus, since $\mathbb{C}$ is an algebraically closed field, it is divisible by a linear polynomial, say $f(x_1) = (x_1 - \alpha_1)g(x_1)$ in $\mathbb{C}[x_1]$. Thus, $0 = \pi_1(x_1 - \alpha_1)\pi_1(g(x_1))$ in K. Then since K is an integral domain, one of $x_1 - \alpha_1$ or

$g(x_1)$ is an element of $\ker(\pi_1)$. Thus, by induction on the degree of $f(x_1)$, we conclude that $\ker(\pi_1)$ contains a linear polynomial. Therefore, by Proposition 5.26, we conclude that $\ker(\pi_1)$ is a maximal ideal of $\mathbb{C}[x_1]$. □

The following version of Hilbert's Nullstellensatz is not the original version Hilbert proved.

THEOREM 5.6. *(**Hilbert's Nullstellensatz**)*

(1) *The maximal ideals of the polynomial ring*

$$\mathbb{C}[x_1, ..., x_n]$$

are in a one-to-one correspondence with points of the complex n-dimensional space $\mathbb{C}^n$.

(2) *A point* $\alpha = (\alpha_1, ..., \alpha_n)$ *in* $\mathbb{C}^n$ *corresponds to the kernel of the evaluation homomorphism* ε_a: $\mathbb{C}[x_1, ..., x_n] \longrightarrow \mathbb{C}$ *defined by*

$$\varepsilon_\alpha(f(x_1, ..., x_n)) = f(\alpha_1, ..., \alpha_n).$$

(3) *The kernel* M_α *of this map is the ideal generated by the linear polynomials*

$$x_1 - \alpha_1, x_2 - \alpha_2, ..., x_n - \alpha_n.$$

PROOF. Let $\alpha = (\alpha_1, ..., \alpha_n)$ be a point in $\mathbb{C}^n$, and let M_α be the kernel of the evaluation homomorphism ε_α: $\mathbb{C}[x_1, ..., x_n] \longrightarrow \mathbb{C}$. Since this homomorphism is surjective and $\mathbb{C}$ is a field, M_α is a maximal ideal of $\mathbb{C}[x_1, ..., x_n]$.

Now, in order to show that M_α is generated by the linear polynomials, $x_1 - \alpha_1, ..., x_n - \alpha_n$, we first observe that we can always expand a polynomial $f(x) \in \mathbb{C}[x_1, ..., x_n]$ in powers of $x_1 - \alpha_1, ..., x_n - \alpha_n$,

$$f(x) = f(\alpha) + \sum_i c_i(x_i - \alpha_i) + \sum_{i,j} c_{ij}(x_i - \alpha_i)(x_j - \alpha_j) + \cdots.$$

The existence of such an expression of a finite sum can be derived by first substituting $x = \alpha + u$ into $f(x)$, and expanding in powers of the variables $u = (u_1, ..., u_n)$, and then substituting $u = x - \alpha$ back into the result. Thus, every term on the right side except $f(\alpha)$ is divisible by at least one of the linear polynomials $(x_i - \alpha_i)$. Thus, if $f(x) \in \ker(\varepsilon_\alpha)$, then $f(\alpha) = 0$ so that $f(x)$ is an element of the ideal generated by these linear polynomials $x_i - \alpha_i$. Since we have already verified that $\ker(\varepsilon_\alpha)$ is a maximal ideal, it must be the same as the ideal $(x_1 - \alpha_1, ..., x_n - \alpha_n)$:

$$\ker(\varepsilon_\alpha) = M_\alpha = (x_1 - \alpha_1, ..., x_n - \alpha_n).$$

This proves part (2) and part (3). It also proves half of part (1), namely,

$$\alpha \longmapsto M_\alpha = (x_1 - \alpha_1, ..., x_n - \alpha_n)$$

is one-to-one. Thus to finish the whole proof, we are left to to prove that every maximal ideal M of $\mathbb{C}[x_1, ..., x_n]$ is equal to M_α for some point $\alpha \in \mathbb{C}^n$.

Let M be a maximal ideal of $\mathbb{C}[x_1, ..., x_n]$. Then, since the bottom field $\mathbb{C}$ is algebraically closed, by Corollary 5.16, we already know that the kernel $\ker(\pi_1)$ of π_1 is either 0 or else it is a maximal ideal of $\mathbb{C}[x_1]$,

$$\pi_1\colon \mathbb{C}[x_1] \overset{i}{\hookrightarrow} \mathbb{C}[x_1, ..., x_n) \overset{\pi}{\longrightarrow} K \ (= \mathbb{C}[x_1, ..., x_n]/M), \qquad \pi_1 = \pi \circ i.$$

CLAIM 5.2. $\ker(\pi_1)$ *can not be the zero ideal.*

Suppose that we have proved that $\ker(\pi_1)$ can not be the zero ideal, then the maximal ideal $\ker(\pi_1) = (x_1 - \alpha_1)$ for some $\alpha_1 \in \mathbb{C}$, and hence working with another variable x_i we see that the maximal ideal M contains linear polynomials $x_i - \alpha_i$ for $i = 1, ..., n$. Thus, we must have $M = (x_1 - \alpha_1, ..., x_n - \alpha_n)$ for some $\alpha = (\alpha_1, ..., \alpha_n) \in \mathbb{C}^n$. Then since we have already verified that $(x_1 - \alpha_1, ..., x_n - \alpha_n) = \ker\ (\varepsilon_\alpha)$, M is the kernel of an evaluation homomorphism ε_α for some $\alpha = (\alpha_1, ..., \alpha_n) \in \mathbb{C}^n$. Thus, we need to prove the claim.

PROOF OF THE CLAIM. Suppose on the contrary that $\ker(\pi_1) = (0)$. Then π_1 maps $\mathbb{C}[x_1]$ isomorphically to its image, which is a subring of K; and hence we can extend this map to the field $\mathbb{C}(x_1)$ of fractions of $\mathbb{C}[x_1]$ by the universal property of the field of fractions (Proposition 5.22), so that the field K has a subfield which is isomorphic to $\mathbb{C}(x_1)$. Then by Corollary 5.15, we know that $\mathbb{C}(x_1)$ has a linearly independent subset consisting of uncountably many elements; consequently, we have a basis of K consisting of uncountably many elements. But, this is impossible because, since $\mathbb{C}[x_1, ..., x_n]$ is a complex vector space with a countably infinite basis consisting of monomials, the quotient ring K is spanned as a complex vector space by a countable family, consisting of the residues of those monomials. Thus, by Lemma 5.22, K has a basis consisting of countably many linearly independent elements. Then since the dimension of the vector space K can not be both countably infinite and uncountably infinite, we must have $\ker(\pi_1) \neq (0)$. □

This finishes the proof of Hilbert's Nullstellensatz. □

Hilbert's Nullstellensatz provides us with an important connection between Geometry and Algebra. It tells us that the maximal ideals in the polynomial ring $\mathbb{C}[x_1, ..., x_n]$ corresponds to the points in $\mathbb{C}^n$. This correspondence can also be used to relate affine varieties to quotient rings of the polynomial ring:

THEOREM 5.7. *Let V be the affine variety in $\mathbb{C}^n$ defined by a system of polynomial equations,*

$$\begin{cases} f_1 &= 0, \\ f_2 &= 0, \\ \cdots & \\ f_s &= 0. \end{cases},$$

and let $I = (f_1, ..., f_s) \subset \mathbb{C}[x_1, ..., x_n]$ be the ideal generated by these polynomials. Then the maximal ideals of the quotient ring

$$R = \mathbb{C}[x_1, ..., x_n]/I$$

are in one-to-one correspondence with the points of V.

PROOF. We first recall that the maximal ideals of the quotient ring R correspond to those maximal ideals M of $\mathbb{C}[x_1, ..., x_n]$ which contain I. On the other hand, by Hilbert's Nullstellensatz, Theorem 5.6, we know that the maximal ideal M of $\mathbb{C}[x_1, ..., x_n]$ is equal to $M_\alpha = (x_1 - \alpha_1, ..., x_n - \alpha_n)$ for some point $\alpha \in \mathbb{C}^n$. Thus, combining these two results together, we have: $I \subset M_\alpha$ if and only if each $f_i \in M_\alpha = \ker(\varepsilon_\alpha)$ if and only if

$$0 = \varepsilon_\alpha(f_i) = \alpha(f_i) = f_i(\alpha)$$

if and only if $\alpha \in V$. □

DEFINITION 5.43. *The quotient ring R in Theorem 5.7 is called the* ***coordinate ring*** *of the affine variety defined by the polynomial equations $f_1 = f_2 = \cdots = f_s = 0$.*

Theorem 5.7 explains that the algebraic properties of the ring R are closely related with the geometry of V. In principle, all properties of the system of polynomial equations

$$\begin{cases} f_1 &= 0, \\ f_2 &= 0, \\ \cdots & \\ f_s &= 0. \end{cases}$$

are reflected in the structure of the ring $R = \mathbb{C}[x_1, ..., x_n]/(f_1, ..., f_s)$, and the theory of this relationship is called **Algebraic Geometry**.

The following theorem says that the zero ring is the only ring with no maximal ideals.

THEOREM 5.8. *(Krull) Let R be a ring. Every ideal I of R which is not the unit ideal is contained in a maximal ideal.*

This theorem is usually proved using the Axiom of Choice, or equivalently, Zorn's Lemma, so we leave it as an exercise.

COROLLARY 5.17. *The only ring R with no maximal ideals in the zero ring.* □

If we combine Theorem 5.7 and Theorem 5.8 together, we obtain another important corollary:

COROLLARY 5.18. ***(Hilbert's Weak Nullstellensatz)*** *If the system of polynomial equations,*

$$\begin{cases} f_1 &= 0, \\ f_2 &= 0, \\ \cdots & \\ f_s &= 0. \end{cases}$$

over $\mathbb{C}$ has no solution in $\mathbb{C}^n$, then we can express 1 as a linear combination of these f_i's,

$$1 = \sum g_i f_i,$$

with polynomial coefficients in $\mathbb{C}[x_1, ..., x_n]$, and conversely.

That is, $V(f_1, ..., f_s) = \varnothing$, then $(f_1, ..., f_s) = \mathbb{C}[x_1, ..., x_n]$, and conversely.

PROOF. The converse is trivial. Now, if the system has no solution, then Hilbert's Nullstellensatz, Theorem 5.7 says that there is no maximal ideal containing the ideal $I = (f_1, ..., f_s)$; hence by Theorem 5.8, I must be a unit ideal (1). □

REMARK 5.37. *Thus, Hilbert's Weak Nullstellensatz is a generalization of Gauss's Fundamental Theorem for Algebra because it states that* "Every system of polynomial equations that generates an ideal which is smaller than $\mathbb{C}[x_1, ..., x_n]$ has a common solution in $\mathbb{C}^n$". *Thus, Hilbert's Weak Nullstellensatz is also called "**Fundamental Theorem of Algebra for polynomials in several variables**".*

The Nullstellensatz has been reformulated in many apparently different ways, and the one we proved above is not its original form. Here is the original form of Nullstellensatz:

THEOREM 5.9. ***(Hilbert's Strong Nullstellensatz)*** *Let*

$$V = V(f_1, ..., f_s)$$

be the affine variety defined by polynomials $f_1, ..., f_s \in \mathbb{C}[x_1, ..., x_n]$. If a polynomial $g \in \mathbb{C}[x_1, ..., x_n]$ vanishes identically zero on V, then there *exists a positive integer N such that*

$$g^N \in I = (f_1, ..., f_s),$$

and conversely.

PROOF. The converse is trivial since $g^N(a_1, ..., a_n) = 0$ if and only if $g(a_1, ..., a_n) = 0$ holds for any point $(a_1, ..., a_n) \in \mathbb{C}^n$.

We consider the $s+1$ polynomials $f_1, ..., f_s, gy - 1$ in the variables $x_1, ..., x_n$, and y, where $gy = g(x_1, ..., x_n)y$ is a polynomial with one more variable y. Then the system of these polynomials does not have a solution, because if $(\alpha_1, ..., \alpha_n, b)$ is a solution of them, then each $f_i(\alpha_1, ..., \alpha_s) = 0$ and $g(\alpha_1, ..., \alpha_n) = 0$ by hypothesis, and hence $gy - 1 = -1$, which is impossible. Thus, by Corollary 5.18 (Hilbert's Weak Nullstellensatz), we know that they generate the unit ideal (1) of $\mathbb{C}[x_1, ..., x_n, y]$. Thus, we have

$$\begin{aligned} 1 &= \sum_i p_i(x_1, ..., x_n, y) f_i(x_1, ..., x_n) \\ &\quad + q(x_1, ..., x_n, y)(g(x_1, ..., x_n)y - 1), \end{aligned}$$

for some $p_i(x_1, ..., x_n, y)$ and $q(x_1, ..., x_n, y) \in \mathbb{C}[x_1, ..., x_n, y]$. Thus, by substituting $y = 1/g(x_1, ..., x_n)$ into this expression, and clearing the denominators of the resulting expression,

$$1 = \sum_i p_i(x_1, ..., x_n, 1/g(x_1, ..., x_n)) f_i(x_1, ..., x_n)$$

we find that there exists a positive integer N such that

$$g(x_1, ..., x_n)^N = \sum_i h_i(x_1, ..., x_n) f_i(x_1, ..., x_n) \in I,$$

where $h_i(x_1, ..., x_n) = g(x_1, ..., x_n)^N p_i(x_1, ..., x_n, 1/g(x_1, ..., x_n))$. □

REMARK 5.38. *The method we applied in the above proof is usually called **Rabinowitch's trick.***

In fact, Hilbert's Strong Nullstellensatz is equivalent to Hilbert's Weak Nullstellensatz:

THEOREM 5.10. *(**Hilbert's Weak Nullstellensatz**) Let I be an ideal of $\mathbb{C}[x_1, ..., x_n]$. Then $V(I) = \varnothing$ if and only if $I = \mathbb{C}[x_1, ..., x_n]$.*

PROOF. The $\Longleftarrow$ direction it trivial because, the constant polynomial 1 has no solution at all.

Conversely, suppose that I is a proper ideal of $\mathbb{C}[x_1, ..., x_n]$, that is, $I \subsetneq \mathbb{C}[x_1, ..., x_n]$. Then it is contained in a maximal ideal M (Theorem 5.8). Then by Theorem 5.6 (Hilbert's Nullstellensatz),

$$M = (x_1 - \alpha_1, ..., x_n - \alpha_n)$$

for some point $(\alpha_1, ..., \alpha_n) \in \mathbb{C}^n$; hence the point

$$(\alpha_1, ..., \alpha_n) \subseteq V(M) \subseteq V(I).$$

That is, $V(I) \neq \varnothing$. □

Another way of describing Hilbert's Strong Nullstellensatz is:

DEFINITION 5.44. *Let I be an ideal of a Noetherian ring R.*

(1) *Then the set defined by*

$$\sqrt{I} = \{r \in R : r^N \in I \text{ for some } N \geq 1\},$$

*is called the **radical** of I.*

(2) *If $\sqrt{I} = I$ holds, then I is called a **radical ideal**.*

COROLLARY 5.19. *Let I be an ideal of a Noetherian ring R. Then radical $\sqrt{I}$ of I is an ideal of R containing I.*

PROOF. We leave its proof as an exercise. □

COROLLARY 5.20. *Let V be an affine variety. Then $I(V)$, i.e., the set of those polynomial functions* vanishing on V, *is a radical ideal.*

PROOF. We leave its proof as an exercise □

COROLLARY 5.21. *Let I be an ideal of the polynomial ring $\mathbb{C}[x_1, ..., x_n]$ over the field $\mathbb{C}$ of complex numbers. Then*

$$I(V(I)) = \sqrt{I}.$$

REMARK 5.39. *Thus, we have proved both Weak and Strong Versions of Hilbert's Nullstellensatz when the bottom field is* $\mathbb{C}$. *When the bottom field* k *is an algebraically closed field in general, we still have the Hilbert's Nullstellensatz and one may find its proof in, for instance, Hungerford's Book, Algebra.*

Before we move into the next Chapter, we want to closely examine what's going on when we solve a system of linear equations in several variables. The following example actually suggests a few very important notions which were nevertheless neither noticed nor well-understood until 1965.

EXAMPLE 5.40. *Let* $f_1 = y-z$, $f_2 = x+2y+3z$, *and* $f_3 = 3x-4y+2z$ *be linear polynomials in* $\mathbb{Q}[x]$. *We consider the ideal* $I = (f_1, f_2, f_3)$ *and the variety* $V(f_1, f_2, f_3)$, *i.e., the solution set of the system of linear equations:*

$$\begin{cases} y - z &= 0, \\ x + 2y + 3z &= 0, \\ 3x - 4y + 2z &= 0. \end{cases}$$

Applying elementary row operations to the following matrix we solve the system:

$$\begin{bmatrix} 0 & 1 & -1 \\ 1 & 2 & 3 \\ 3 & -4 & 2 \end{bmatrix} \longrightarrow \begin{bmatrix} 0 & 1 & -1 \\ 1 & 2 & 3 \\ 0 & -10 & -7 \end{bmatrix} \longrightarrow \begin{bmatrix} 1 & 2 & 3 \\ 0 & 1 & -1 \\ 0 & 0 & -17 \end{bmatrix}.$$

We recall that any one of the three elementary row operations never change the solution set at all. That is, the solution set must be invariant under these elementary row operations. In this example, we thus obtain: $V(f_1, f_2, f_3) = V(f_1, f_2, -17z)$. *In fact, we have more than that:*

$$(f_1, f_2, f_3),$$
$$(f_1, f_2, f_3 - 3f_2),$$
$$(f_2, f_1, f_3 - 3f_2 + 10f_1),$$

where $f_3 - 3f_2 + 10f_1 = -17z$. *We note that they are the same ideal* I, *but its generators are transformed by elementary row operations to the next ones in such a way that we can easily figure out its solution set!*

REMARK 5.40. *There are a few important but usually not-mentioned notions in the business of elementary row operations for solving a system of linear equations.*

First of all, we psychologically think that x *is greater than* y *is greater than* z; *and then try to eliminate the term with* x *in* f_3; *then try to eliminate the term with* y *in* f_3. *More precisely, we try to eliminate the leading term of* f_3 *by the leading term of* f_2; *then we try to eliminate the leading term of the resulting polynomial,* $f_3 - 3f_2$ *by the leading term of* f_1.

Secondly, we stopped because the resulting term $-17z = f_3 - 3f_2 + 10f_1$ *is "small enough" so that none of the leading terms of either* f_1 *or* f_2 *can eliminate, or reduce further, the resulting term,* $-17z$.

Exercises for Chapter 5.9

EXERCISE 5.56. *Prove that every maximal ideal of a commutative ring is a prime ideal.*

EXERCISE 5.57. *Complete the proof of Proposition 5.25.*

EXERCISE 5.58. *Show that the ideal* (x) *is a prime ideal but not a maximal ideal of* $\mathbb{Z}[x]$.

EXERCISE 5.59. *Find all subrings of* $\mathbb{Q}$.

CHAPTER 6

Factorization

1. Introduction

In 1637, Fermat (1601 – 1665) stated that the Diophantine equation

$$x^n + y^n = z^n,$$

with integer $n > 2$ has no positive integer solutions x, y, z. For small numbers $n = 3$, or even $n \leq 125,000$ the statement has been proved. But trying to adapt the method of proof that works for prime $n = 3$ to the general case, Kummer (1810 – 1893) thought he found a proof, but Dirichlet (1805 – 1859) observed that at one point the argument Kummer gave had a gap. It later turned out that the difficulty consist in the fact the cyclotomic fields in which we are led to operate are not a unique factorization domain if prime $n > 19$. Until that time, most people thought that the uniqueness of factorization of natural numbers into primes is so obvious that a formal proof is virtually a waste of time. But, Kummer and Dirichlet recognized that in an only slightly more general setting the statement of essential uniqueness of factorizations is actually false. More historical accounts in this direction are, for instance, in E. Grosswald's book, *Topics from the Theory of Numbers*, 2nd Edition.

We first recall that multiplication of two numbers is a function as well as addition of two number is so. Thus, whoever computes them *in that direction* the result of such computations on two numbers respectively are the same.

But, *the reverse computation* in such a situation is not always very happy. Here is our first example:

EXAMPLE 6.1. *In the ring* $\mathbb{Z}[\sqrt{-5}]$, *which is an extended integral domain of* $\mathbb{Z}$, *we have*

$$6 = 2 \cdot 3 = (1 + \sqrt{-5})(1 - \sqrt{-5}).$$

We also have

$$3^2 = 3 \cdot 3 = (2 + \sqrt{-5}) \cdot (2 - \sqrt{-5}).$$

This example tells us, first of all, that the number 6 has two different factorizations and hence the ring $\mathbb{Z}[\sqrt{-5}]$ is not a *UFD* (Unique Factorization Domain). We recall that $\mathbb{Z}$ is a *PID* (Principal Ideal Domain), and hence it is automatically a *UFD*. But, its extended ring $\mathbb{Z}[\sqrt{-5}]$ is not a *UFD*; hence it is not a *PID*.

In other words, this example says that, depending on where we are working, the "smaller ones", even though you have figured them out properly, whose multiplication-results in the given number 6, are *not* uniquely determined.

We note that the above example is not the only such example where the uniqueness of factorization fails. In fact, (although we are not going to prove it in this book) one can prove that

THEOREM 6.1. *Let* d *be a negative integer. The ring* $\mathbb{Z}[\sqrt{d}]$ *of integers (in the imaginary quadratic field* $\mathbb{Q}[\sqrt{d}]$*) is a Unique Factorization Domain (UFD) if and only if* $d = -1, -2, -3, -7, -11, -19, -43, -67,$ *or* -163. □

REMARK 6.1. *Here is some historical remark: Gauss proved the* $\Leftarrow$ *direction of the above theorem. He also conjectured that there were no others. Thus much more difficult part of the theorem above was finally proved by Baker and Stark in* 1966, *after the problem had been worked on for more than* 150 *years.*

On the other hand, the following example should still confuse at least some of you:

EXAMPLE 6.2. *The number* 5 *in* $\mathbb{N}$ *or* $\mathbb{Z}$ *is prime and hence it is irreducible, but if we enter the ring* $\mathbb{Z}[i]$ *of Gauss integers, which by definition, consists of those numbers of the form* $a + bi$*, where* $a, b \in \mathbb{Z}$*, we have apparently* 4 *different factorization of the number* 5*:*

$$5 = (2+i)(2-i) = (-2-i)(-2+i) = (1-2i)(1+2i) = (-1+2i)(-1-2i),$$

where $i = \sqrt{-1}$*, that is,* i *is a number whose square is equal to* -1*. Thus, the number* 5 *is not prime and hence not irreducible anymore there. Later, we will learn that, in a UFD, irreducible elements are prime elements, and conversely.*

On the other hand, we know that it is a UFD, for instance, by Theorem 6.1 above. But, since we have exactly 4 *units,* $\pm 1, \pm i$ *in* $\mathbb{Z}[i]$ *, these apparently different factorizations*

$$(2+i)(2-i),\ (-2-i)(-2+i),\ (1-2i)(1+2i),\ (-1+2i)(-1-2i)$$

are essentially the same, i.e., equivalent. This is because the 4 *associates of* $2+i$ *are*

$$2+i,\ -2-i,\ -1+2i,\ 1-2i,$$

and the 4 *associates of* $2-i$ *are*

$$2-i,\ -2+i,\ 1+2i,\ -1-2i.$$

Therefore, we still have only one factorization of the number 5 *in ring* $\mathbb{Z}[i]$ *of Gauss integers.*

REMARK 6.2. *We recall that a field* k *is, by definition, an integral domain where every non-zero element is invertible. Thus, no meaningful factorization of an element of a field* k *is ever possible. From now on we assume that all our rings are commutative rings.*

We have seen that the uniqueness of factorization of numbers fails when we extend $\mathbb{Z}$ to a certain larger integral domain $\mathbb{Z}[\sqrt{d}]$.

This kind of problem was a main difficulty of Kummer when he thought that he found a proof of Fermat's Last Theorem. In 1983 Faltings proved that there are only finitely many integer solutions for Fermat's Last Theorem, for any given value of n. It is now known that it is finally (and completely) proved by an English mathematician, Andrew Wiles in his famous paper, *Modular elliptic curves and Fermat's last theorem,* Pages 443-551, Annals of Mathematics 141(3), May 1995. It thus took about 350 years for its complete proof and it is one of the major triumphs of mathematics of the 20th century.

We think this is sufficient enough for an introduction to this Chapter. In what follows, we will study factorizations in a more systematic or axiomatic manner.

Exercises for Chapter 6.1

EXERCISE 6.1. *Prove that* $\pm 1, \pm i$ *are the only units in* $\mathbb{Z}[i]$.

2. Factorization of Polynomials in One Variable

The main purpose of this chapter is a study of division in rings. Because it is modeled on properties of the ring $\mathbb{Z}$ of integers, we need to review these properties. The results in this particular section are a continuation of those in the second-half of the last chapter.

REMARK 6.3. *We note that as long as factorization of polynomials are concerned, treatment of those polynomials in one variable is completely different from treatment of those polynomials in more than one variable. This is mainly because a polynomial ring* $k[x]$ *in one variable over a field* k *is a UFD that is also a PID, whereas a polynomial ring* $k[x, y]$ *in two variables over a field* k *is a UFD but is not a PID.*

In this section, we will study polynomials in one variable only, because a proper treatment of polynomials in more than one variables does require the theory of Groebner bases, discovered and developed by Bruno Buchberger (1942 -) in late 20th *century.*

We start our discussion with an elementary school arithmetic stuff:

THEOREM 6.2. *(Division with Remainder) If a, b are two distinct natural numbers and $a \neq 0$, then there exist unique integers q, r such that*

$$b = aq + r, \tag{6.1}$$

with $0 \leq r < a$. □

Thus, whoever computes the quotient q and the remainder r, the results are going to be the same. Moreover, we have an **algorithm** for actually computing q and r for a given set of inputs: b and $a \neq 0$, which we learned it even when we are in an elementary school.

One immediate consequence of this theorem, which we already proved in the previous Chapter, is:

COROLLARY 6.1. *$\mathbb{Z}$ is a PID.* □

We note that given non-zero natural numbers a, b, the set defined by

$$a\mathbb{Z} + b\mathbb{Z} := \{ar + bs : r, s \in \mathbb{Z}\}$$

is again an ideal in $\mathbb{Z}$, the smallest ideal containing both principal ideals $a\mathbb{Z}$ and $b\mathbb{Z}$, and by this Corollary it is generated by a single element, say d:

$$d\mathbb{Z} = a\mathbb{Z} + b\mathbb{Z}.$$

The immediate question one can think of is: how to compute d explicitly.

LEMMA 6.1. $d = \gcd(a, b)$.

PROOF. Certainly, $d = ar + bs$ for some $r, s \in \mathbb{Z}$. Also, d clearly divides both a and b since $a = a \cdot 1 + b \cdot 0$ for instance. Now, if h divides both a and b, then a and b are elements of the ideal $h\mathbb{Z}$; hence $d\mathbb{Z} = a\mathbb{Z} + b\mathbb{Z} \subset h\mathbb{Z}$; hence h does divide d.This exactly means that d is the greatest common divisor of a and b. □

COROLLARY 6.2. $\gcd(a, b) = ar + bs$ *for some $r, s \in \mathbb{Z}$.* □

REMARK 6.4. *We note that the above treatments do not tell us how we compute* $\gcd(a, b)$, *r and s. To compute* $\gcd(a, b), r$, *and s, we apply the* *very well-known* ***division algorithm of Euclid*** *of linear time efficiency, which is known to us since about BC* 300. *Here is the division algorithm of Euclid for* $\gcd(a, b)$.

Algorithm 1 (Euclid's algorithm for $\gcd(a, b)$) We may assume that $q_0 = b > a_1 = a \neq 0$. If b is a multiple of $a \neq 0$, then $a = \gcd(a, b)$. Otherwise, we continuously apply the above theorem to obtain:

$$\begin{aligned}
b &= q_1 a + r_1, \text{ with } r_1 > 0, \\
a &= q_2 a_1 + r_2, \text{ with } a_1 = r_1 \text{ and } r_2 > 0, \\
a_1 &= q_3 a_2 + r_3, \text{ with } a_2 = r_2 \text{ and } r_3 > 0, \\
& \quad \cdots\cdots\cdots \\
a_{k-2} &= q_k a_{k-1} + r_k, \text{ with } a_{k-1} = r_{k-1} \text{ and } r_k > 0, \\
a_{k-1} &= q_{k+1} a_k + 0, \text{ with } a_k = r_k > 0 \text{ but } r_{k+1} = 0.
\end{aligned}$$

Then $r_1 > r_2 > ... > r_k > 0$ and hence $\gcd(a, b) = .\gcd(r_1, r_2) = \gcd(r_2, r_3) = ... = \gcd(r_k, 0) = r_k$. □

This algorithm does work since

$$\gcd(b, a) = \gcd(b - aq, a) = \gcd(r, a) = \gcd(a, r),$$

and by rewriting these equations backwards starting from $\gcd(a, b)$, we can find those integers r and s in $\gcd(a, b) = ar + bs$. We note that these r and s are not uniquely determined.

EXAMPLE 6.3. *We know by inspection that* $\gcd(6, 29) = 1$. *On the other hand,*

$$\begin{aligned}
29 &= 4 \cdot 6 + 5, \\
6 &= 1 \cdot 5 + 1, \\
5 &= 5 \cdot 1 + 0.
\end{aligned}$$

Thus, $1 = 6 - 1 \cdot 5 = 6 - 1 \cdot (29 - 4 \cdot 6) = 5 \cdot 6 + (-1) \cdot 29$; *hence* $r = 5$ *and* $s = -1$. *Obviously, these* r *and* s *are not determined uniquely:*

$1 = 5 \cdot 6 + \{(-29) \cdot 6 + 29 \cdot 6\} + (-1) \cdot 29 = (-24) \cdot 6 + 28 \cdot 29$, *where* $6 \cdot 29 = \mathrm{lcm}(6, 29)$, *etc.*

LEMMA 6.2. *Let p be a prime number and let a, b be numbers. If p divides their product ab, then p divides at least one of them.*

PROOF. Suppose that the prime p does not divide a. Then $\gcd(p, a) = 1$; hence by the above corollary we have

$$1 = pr + as$$

for some $r, s \in \mathbb{Z}$. Multiplying both sides by b we see that p does divide b. □

THEOREM 6.3. ***(Fundamental Theorem of Arithmetic)*** *Every non-zero integer a can be written as a product*

$$a = cp_1 \cdots p_k, \tag{6.2}$$

where $c = \pm 1$, the p_i are prime numbers, and $k \geq 0$. This expression is unique except for the ordering of the prime factors p_i.

PROOF. First, we prove that a prime factorization exists: To prove this, it is enough to consider the case that a is greater than 1. By induction on a, we may assume that the existence is proved for all positive integers $b < a$. Either a is prime, in which case the product has one factor, or there is a proper divisor $b \neq a$. Then $a = bb'$, where both b and b' are smaller than a, and hence by induction hypothesis they can be factored into primes. Setting their factorization side by side gives a factorization of a.

Second, we prove that the factorization is unique: For this we suppose that

$$\pm p_1 \cdots p_n = \pm q_1 \cdots q_m.$$

The signs should certainly agree. Then we apply the above lemma with $p = p_1$: Since p_1 divides the product $q_1 \cdots q_m$ it divides some q_i, say q_1. Since q_1 is also prime, we must have that $p_1 = q_1$. We cancel p_1 from both sides and we proceed by induction, noting that 1 is never equal to a product of primes. □

REMARK 6.5. *Although the above factorization is unique up to units* $c = \pm 1$*, it is usually very difficult to compute those prime factors* p_i *when* a *is large. Moreover, the above proof does not tell us how one can compute those prime factors* p_i *efficiently.*

As we mentioned earlier, the structure of the ring $\mathbb{Z}$ of integers is closely analogous to that of a polynomial ring $k[x]$ in one variable x over a field k. Whenever a property of one of these two rings is derived, we should try to find an analogous property of the other.

EXAMPLE 6.4. *We have already seen that both* $\mathbb{Z}$ *and* $k[x]$ *are a PID in the previous chapter.*

DEFINITION 6.1. *A polynomial* $p(x)$ *with coefficients in a field* k *is called* ***irreducible*** *over* k *if it is not constant and if its only divisors of lower degree in* $k[x]$ *are constants.*

This means that the only way that p can be written as a product of two polynomials is $p = cp_1$ where c is a constant, i.e., an element of k, and p_1 is a constant multiple of p. The irreducible polynomials are analogous to prime integers, i.e., prime numbers. Since our polynomials have their coefficients in the field k, it is customary to normalize them by factoring out their leading coefficients, i.e., the coefficient of the monomial of the polynomial with highest degree, so that they become *monic.*

The proof of the following theorem is similar to the proof of the analogous statements for the ring of integers.

THEOREM 6.4. *Let* k *be a field and let* $k[x]$ *denote the polynomial ring in one variable* x *over a field* k.

- **a:** *If two polynomials* f, g *have no common non-constant factor, then there are polynomials* $r, s \in k[x]$ *such that* $fr + gs = 1$.
- **b:** *If an irreducible polynomial* $p \in k[x]$ *divides a product* fg*, then* p *divides one of the factors,* f *or* g.

c: *Every non-zero polynomial $f \in k[x]$ can be written as a product*

$$f = cp_1 \cdots p_k,$$

where c is a non-zero constant, the p_i are monic irreducible polynomials in $k[x]$, and $k \geq 0$. The factorization is unique, except for the ordering of the terms. □

REMARK 6.6. *To compute the r and s in part a of the above theorem, we simply apply the algorithm of Euclid for* $\gcd(f, g)$*, which is exactly the same as the above Algorithm, except for having* $\deg(r_1) > \ldots \deg(r_{k-1}) > 0$*, instead of having* $_1 > \ldots > r_{k-1}$ *there.*

The constant c in the third part of this theorem is analogous to the factor ± 1 in (6.2). These are units in their respective rings. The unit factors are there because we *normalized* primes to be positive, and irreducible polynomials to be monic. We can allow negative primes or non-monic irreducible polynomials if we wish. The unit factor can then be absorbed, if $k > 0$. But, this complicates the statement of uniqueness slightly.

The next examples shows that factorizations of polynomials in one variable over either the field $\mathbb{C}$ of complex numbers or the field $\mathbb{R}$ of real number are fairly easy.

EXAMPLE 6.5. *Over the field $\mathbb{C}$ of complex numbers, thanks to Gauss' Fundamental Theorem of Algebra, Theorem 1.1, every polynomial of positive degree has a complex root α and hence has a divisor of the form $x - \alpha$. Therefore, the irreducible polynomials over $\mathbb{C}$ are linear, and the irreducible factorization of a polynomial over $\mathbb{C}$ has the form*

$$f(x) = c(x - \alpha_1) \cdots (x - \alpha_n), \tag{6.3}$$

where α_i are the roots of $f(x)$, repeated as necessary. Thus, the uniqueness of this factorization is not surprising.

REMARK 6.7. *Over the field $\mathbb{R}$ of real numbers, there are only two classes of irreducible polynomials: linear polynomials and irreducible quadratic polynomials. We recall that a real quadratic polynomial x^2+bx+c is* *irreducible over* $\mathbb{R}$ *if and only if its discriminant* $b^2 - 4c$ *is negative, in which case it has a pair of complex conjugate roots.* The fact that every irreducible polynomial over the complex numbers is linear implies that no higher-degree polynomial is irreducible over the reals. *This goes as follows: Suppose that a polynomial* $f(x)$ *has real coefficients* a_i *and that* α *is a complex, non-real root of* $f(x)$. *Then the complex conjugate* $\bar{\alpha}$ *is different from* α *and is also a root. For, since it is a real polynomial, the its coefficients* a_i *satisfy* $a_i = \bar{a}_i$, *and hence*

$$f(\bar{\alpha}) = a_n\bar{\alpha}^n + ... + a_1\alpha + a_0 = \bar{a}_n\bar{\alpha}^n + ... + \bar{a}_1\alpha + \bar{a}_0 = \overline{f(\alpha)} = \bar{0} = 0.$$

Thus the quadratic polynomial $g(x) := (x-\alpha)(x-\bar{\alpha}) = x^2-(\alpha+\bar{\alpha})x+\alpha\bar{\alpha}$ *has real coefficients* $-(\alpha+\bar{\alpha})$ *and* $\alpha\bar{\alpha}$, *and moreover, both of its linear factors appear on the right side of the complex factorization (6.3) of* $f(x)$. *Thus,* $g(x)$ *divides* $f(x)$. *This means that the factorization of* $f(x)$ *into irreducible real polynomials is obtained by grouping conjugate pairs in the complex factorization of* $f(x)$.

REMARK 6.8. *But, factorization of polynomials in one variable is more complicated for* polynomials with rational coefficients *than for real or complex coefficients, because there exist irreducible polynomials in* $\mathbb{Q}[x]$ *of arbitrary degree.*

EXAMPLE 6.6. *For any* $n > 0$, *the polynomial* $x^n - 3x^{n-1} + 3$ *is irreducible over* $\mathbb{Q}$, *by Eisenstein's Criterion for irreducibility of polynomials with rational coefficients. We will later study his criterion in detail. (See Proposition 6.11).*

Neither the form of the irreducible factorization nor its uniqueness is intuitively clear for rational polynomials in one variable.

REMARK 6.9. *Because of several hypotheses of Eisenstein's Criterion, it helps us to determine only whether rational polynomials special type are irreducible or not, as we see in this Example. That is, for general rational polynomials Eisenstein's Criterion is not useful at all. The next trial we* *are going to do is to study of factorizations in $\mathbb{Z}[x]$, which in fact provides the main idea for the proof of Eisenstein Criterion, Proposition 6.11.*

The main reason why we want to move into $\mathbb{Z}[x]$ may be the following. In there, we have a natural ring homomorphism $\mathbb{Z}[x] \longrightarrow \mathbb{Z}_p[x]$, defined by

$$f(x) = a_n x^n + \cdots + a_0 \longmapsto \bar{f}(x) = \bar{a}_n x^n + \cdots + \cdot \bar{a}_0,$$

where $\mathbb{Z}_p := \mathbb{Z}/p\mathbb{Z}$ is a field for each prime number p; hence the cardinality of $\mathbb{Z}_p$ is p, i.e., finite; hence in $\mathbb{Z}_p[x]$, there are only finitely many polynomials of the same degree. *Here, $\bar{a}_i$ denotes the residue class of $a_i \bmod p$. We call $\bar{f}(x)$ the* **residue** *of $f(x) \bmod p$. Clearly, this ring homomorphism is induced by the projection: $\mathbb{Z} \longrightarrow \mathbb{Z}_p$ and its kernel is $p\mathbb{Z}[x]$.*

This ring homomorphism sometimes makes our life quite easier because it can be used to determine whether a given primitive *polynomial $f(x)$ is an irreducible polynomial or not. (Cf. Definition 6.5). That is, if there exists a prime integer p, not dividing a_n, such that the residue $\bar{f}(x) \bmod p$ is irreducible, then so is $f(x)$. Therefore, for a decision for smaller degree primitive polynomials to be irreducible or not, this homomorphism is very useful, as we shall explicitly see it in the next section.*

But it turned out that a close examination of factorization of rational polynomials requires us to introduce more fine terminologies and more abstract concepts in order to analyze the problem more clearly. This is basically done in the next section. Then, thanks to Gauss, we can prove that $\mathbb{Z}[x]$ is a UFD, and other important results for later usage. This is treated in the section, called Gauss's Lemma below.

For future reference, we have the following

PROPOSITION 6.1. *A polynomial $f(x)$ of degree n with coefficients in an integral domain R has at most n roots in R.*

PROOF. We know that $\alpha \in R$ is a root of f if and only if $x - \alpha$ divides $f(x)$. If it is the case, we can write $f(x) = (x - \alpha)q(x)$, where $q(x)$ is a polynomial of degree $n - 1$. If β is another root of $f(x)$, then $f(\beta) =$

$(\beta - \alpha)q(\beta) = 0$. Since R is an integral domain, it has no zero divisors. Thus, one of the two elements $(\beta - \alpha), q(\beta)$ must be zero. Thus, either $\beta = \alpha$, or $q(\beta) = 0$. Then by induction hypothesis, we may assume that $q(x)$ has at most $n - 1$ roots in R. Then there are at most n possibilities for β. □

REMARK 6.10. *The fact that R is an integral domain is crucial to Proposition 6.1, as the following example shows.*

EXAMPLE 6.7. *Let R be the quotient ring $\mathbb{Z}/8\mathbb{Z}$. Then in the polynomial ring $R[x]$, we have*

$$\begin{aligned} x^2 - 1 &= (x+1)(x-1) = (x+3)(x-3) \\ &= (x+5)(x-5) = (x+7)(x-7). \end{aligned}$$

Thus, the polynomial $x^2 - 1$ has four roots $\bmod 8$, *and its factorization into irreducible polynomials is not unique.*

Exercises for Chapter 6.2

EXERCISE 6.2. *Let a, b be positive integers whose sum is a prime p. Prove that their greatest common divisor is* 1.

EXERCISE 6.3. *Prove that if n is a positive integer which is not a square of an integer, then $\sqrt{n}$ is not a rational number. For instance, $\sqrt{2}$ is not a rational number.*

EXERCISE 6.4. *Compute the greatest common divisor of the following polynomials:*

$$x^3 - 6x^2 + x + 4,$$

$$x^5 - 6x + 1.$$

EXERCISE 6.5. *Prove that there are infinitely many prime integers.*

3. *UFD*'s and *PID*'s

In the previous section, we have seen that factorization of polynomials in one variable is analogous to factorization of integers. Now, it is natural to ask whether other rings can have such properties. Relatively few such rings exist, but the ring $\mathbb{Z}[i]$ of Gauss integers is one interesting example. Before we study the so-called Gauss' Lemma, which is definitely needed in the study of irreducibility of polynomials in $\mathbb{Z}[x]$, we need to introduce more terminologies for a better study of factorization.

From now on, we assume that the given ring R is an integral domain, so that the Cancellation Law is available.

REMARK 6.11. *In order to describe factorization properly, we need the following definitions. Since we are too much familiar with the ring $\mathbb{Z}$ of integers, we have a tendency of identifying an irreducible element with a prime element, but when we are in an extended integral domain of $\mathbb{Z}$ it is not always the case, as we have already seen in examples in the previous section.*

DEFINITION 6.2. *Let R be an integral domain.*

(1) *Let $a, b \in R$. We say that an element a **divides** another element b (sometimes abbreviated $a|b$) if $b = aq$ for some $q \in R$.*
(2) *An element a in R is a **proper divisor** of another element $b \in R$ if $b = aq$ for some $q \in R$ and neither a nor q is a unit (i.e. an invertible element).*
(3) *A non-zero element in R is called **irreducible** if it is not a unit (i.e., invertible) and it has no proper divisor.*
(4) *Two elements $a, a' \in R$ are called **associates** if each divides the other.*
(5) *An element a in R is called **prime** if a is not zero and not a unit, and if a divides a product bc of two elements b, c of R then it divides one of the factors, b or c.*

It is easily seen that a, a' are associates if and only if they differ by a unit factor, that is, $a' = ua$ for some unit $u \in R$.

We also note that by definition, the only divisors of an irreducible elements are units and its associates.

LEMMA 6.3. *A prime element in an integral domain R is irreducible.*

PROOF. (By contradiction) Let $a \in R$ be a prime element. If it were not an irreducible element, then $a = bc$ for some non-units, i.e., non-invertible elements, b and c in R. Then since a is prime and since we obviously have $a|a$, a divides one of its factors, b or c. Say a divides b, so $b = ad$ for some $d \in R$. Then $a = bc = adc$; hence $1 = dc$ by the Cancellation Law, and hence c is invertible. But, this is a contradiction to our choice of b and c. Therefore, the prime a must be irreducible. □

REMARK 6.12. *The converse of this lemma does not hold in general, unless R is a UFD (Unique Factorization Domain). (See Example 6.9 below).*

The concepts of divisor, unit, and associate can be interpreted in terms of the *principal ideals* generated by any one of these elements.

REMARK 6.13. *This approach is definitely needed since it provides us a means to compare which one is bigger than or smaller than the other. In $\mathbb{Z}$, we obviously have a total ordering on it so that any proper factor of a number has a "smaller" size.*

DEFINITION 6.3. *An ideal I in a ring R is called* **principal** *if it is generated by a single element:*

$$I = (a) := \{ra : r \in R\}.$$

Then one can easily check that

(1) u is a unit if and only if $(u) = (1)$.
(2) a and a' are associates if and only if $(a) = (a')$.
(3) a divides b if and only if $(a) \supseteq (b)$.
(4) a is a proper divisor of b if and only if $(1) \supset (a) \supset (b)$, where the inclusion $\supset$ is a proper inclusion.

Since we are looking for establishing a theorem that is analogous to the Fundamental Theorem of Arithmetic in an integral domain R (Theorem 6.3 above), we may divide the statement of the theorem into two parts. First, a given element a must be a product of irreducible elements, and second, this product must be essentially unique.

Since the second part usually causes us more trouble than the first part, we first consider the first part in detail. For this, we assume that our element a is neither zero nor a unit; otherwise, we have no hope of writing it as a product of irreducible elements. Then we attempt to factor a, proceeding as follows: If a is irreducible itself, we are done. If not, then a has a proper factor, so it decomposes in some way as a product, $a = a_1b_1$, where neither a_1 nor b_1 is a unit. We continue factoring a_1 and b_1 if possible, and we hope that this procedure terminates; in other words, we hope that after a finite number of steps all the factors at our hands are irreducibles. The condition that this procedure always terminates has a neat description in terms of principal ideals:

PROPOSITION 6.2. *Let R be an integral domain. Then the following two statements are equivalent:*

- **a:** *For every non-zero element a of R which is not a unit, the process of factoring a terminates after finitely many steps and results in a factorization $a = b_1b_2 \cdots b_k$ of a into irreducible elements of R.*
- **b:** *R does not contain an infinite increasing chain of principal ideals*

$$(a_1) \subset (a_2) \subset (a_3) \subset ...,$$

where the inclusion $\subset$ is proper.

PROOF. Suppose that R contains an infinite increasing sequence $(a_1) \subset (a_2) \subset (a_3) \subset ...$. Then $(a_n) \subset (1)$ for every n because $(a_n) \subset (a_{n+1}) \subseteq (1)$. Since $(a_{n-1}) \subset (a_n)$, a_n is a proper divisor of a_{n-1}, say $a_{n-1} = a_nb_n$, where both a_n and b_n are not units. This provides a non-terminating sequence of factorizations of a_1: $a_1 = a_2b_2 = a_3b_3b_2 = a_4b_4b_3b_2 = ...$

Conversely, such a sequence of factorizations gives us an increasing chain of ideals. □

Remark 6.14. *The second condition of this proposition is often called the* ***ascending chain condition*** *for principal ideals. We will say, by definition, that* ***existence of factorizations*** *holds in R if the equivalent conditions of the proposition are true.*

Here is an example in which existence of factorizations fails.

Example 6.8. *Let $R = k[x_1, x_2, ...]$, where for $m = 1, 2, ...,$ x_{m+1} is the 2^m-th root of x_1. That is, R is obtained by adjoining all 2^m-th roots of the variable x_1 to the polynomial ring $k[x_1]$, with relations, $x_2^2 = x_1, x_3^2 = x_2$, and so on. Then we can factor the element x_1 indefinitely in this ring, and correspondingly there is an infinite chain $(x_1) \subset (x_2) \subset ...$ of principal ideals.*

It turns out that we need infinitely many generators for a ring to make an example such as the one given above, so we will rarely encounter such rings.

In practice, the second part of the Fundamental Theorem (Theorem 6.3) is the one which gives the most trouble. Factorization into irreducible elements will usually be possible, but it will not be unique.

Units in a ring complicate the statement of uniqueness. It is clear that unit factors should be disregarded, since there is no end to the possibility of adding unit factors in pairs uu^{-1}. For the same reason, *associate* factors should be considered equivalent.

The units in the ring $\mathbb{Z}$ of integers are ± 1, and in this ring it was natural to normalize irreducible element (primes) to be positive; similarly, we may normalize irreducible polynomials in one variable with coefficients in a field k by insisting that they be monic. But, we do not have a reasonable way to normalize elements of an arbitrary integral domain, so we will allow some ambiguity.

It is actually neater to work with *principal ideals* than with elements: Associates generate the same principal ideal. However, it is not too cumbersome to use elements here, and we will stay with them.

DEFINITION 6.4. *We will call an integral domain a* ***Unique Factorization Domain (UFD)*** *if it satisfies the following properties.*

(1) *Existence of factorizations is true for R. In other words, the process of factoring a non-zero element a which is not a unit terminates after finitely many steps and yields a factorization $a = p_1 \cdots p_m$, where each p_i is irreducible.*
(2) *The irreducible factorization of an element is unique in the following sense: If a is factored in two ways into irreducible elements, say $a = p_1 \cdots p_m = q_1 \cdots q_n$, then $m = n$ and with suitable ordering of the factors, p_i is an associate of q_i for each i.*

Thus in the statement of uniqueness, associate factorizations are considered **equivalent**.

Here is an example where uniqueness of factorization fails.

EXAMPLE 6.9. *Let $R = \mathbb{Z}[\sqrt{-5}]$, consisting of all complex numbers of the form, $a + b\sqrt{-5}$, where $a, b \in \mathbb{Z}$. Since this ring is a subring of the field $\mathbb{C}$ of complex numbers, this ring is an integral domain. In this ring, units are ± 1, and the integer 6 has two essentially different factorizations in R:*

$$6 = 2 \cdot 3 = (1 + \sqrt{-5})(1 - \sqrt{-5}).$$

Using the ***norm function*** *$N\colon R \longrightarrow \mathbb{N} \cup \{0\}$ defined by*

$$N(\alpha) = \alpha\bar{\alpha} \ (= |\alpha|^2) = a^2 + 5b^2,$$

where $\alpha = a + b\sqrt{-5}$ with $a, b \in \mathbb{Z}$ and $\bar{\alpha} := a - b\sqrt{-5}$, we can show, for instance, that $1 + \sqrt{-5}$ is irreducible, and it goes as follows:

By definition, this function is clearly multiplicative:

$$N(\alpha\beta) = N(\alpha)N(\beta),$$

and moreover $N(\alpha)$ is a positive integer unless $\alpha = 0$.

We have $N(1+\sqrt{-5}) = 6 = 2 \cdot 3$, *which is in particular a product of prime numbers* 2 *and* 3 *in* $\mathbb{N}$.

Now, if $1+\sqrt{-5}$ *were reducible, then the norm of its proper factor,* $a+b\sqrt{-5}$, *with* $a, b \in \mathbb{Z}$, *must be equal to either the prime number* 2 *or the prime number* 3. *But,* $N(a+b\sqrt{-5}) = a^2 + 5b^2$ *is always strictly bigger than* 3, *unless* $b = 0$. *Consequently, we must have* $b = 0$. *If* $b = 0$, *then its proper factor* a *must be either* $\pm\sqrt{2}$ *or* $\pm\sqrt{3}$, *but both of them are not in* R. *Therefore,* $1+\sqrt{-5}$ *is indeed irreducible.*

By a similar argument, one can show that other factors, $2, 3$, *and* $1-\sqrt{-5}$ *are all irreducible in* R.

Thus, the prime numbers 2 *and* 3 *are not primes in* R *any more.*

For units in R, *we can easily prove that* $a+b\sqrt{-5}$ *is a unit in* R *if and only if* $N(a+b\sqrt{-5})$ $(= a^2+5b^2) = 1$. *That is,* units are those elements in R lying on the unit circle in the complex plane, centered at the origin.

Then by a similar argument as we have done above, we easily see that the only units in R *are* ± 1.

Thus, both 2 *and* 3 *are not associates of* $1 \pm \sqrt{-5}$, *which shows that the two factorizations above are essentially different and hence that* $R = \mathbb{Z}[\sqrt{-5}]$ *is not a unique factorization domain.*

The crucial property of prime integers is that if a prime divides a product, it divides one of the factors, and this property is used as a defining property for an element in an integral domain to be prime (Definition 6.2.)

PROPOSITION 6.3. *Let* R *be an integral domain. Suppose that existence of factorizations holds in* R. *Then* R *is a Unique Factorization Domain if and only if every irreducible element is prime.*

The proof is a simple extension of the arguments used in Lemma 2.9 and Theorem 6.3:

PROOF. If every irreducible is prime, then R is indeed a *UFD*, since existence of factorizations in R is guaranteed by hypothesis. Conversely, suppose that R is a *UFD* and let $a \in R$ be irreducible. To prove that a is prime, we assume that $a|bc$ with $b, c \in R$. Then $bc = ad$ for some $d \in R$. Since R is a *UFD*, we break b, c, and d into their irreducible factors to have:

$$\alpha p_1 \cdots p_k \cdot q_1 \cdots q_l = a \cdot r_1 \cdots r_m,$$

where α is a unit and every other factors are all irreducible. Thus, either one of the p_i's or one of the q_i's must be an associate of the irreducible a; hence $a|b$ or $a|c$. □

COROLLARY 6.3. *In a UFD, every irreducible element is prime.* □

Its proof is given in the proof of Proposition 6.3.

REMARK 6.15. *It is very important to distinguish the two concepts of irreducible element and prime element. They are equivalent in Unique Factorization Domains, but most rings contain irreducible elements which are not prime.*

EXAMPLE 6.10. *In the ring $R = \mathbb{Z}[\sqrt{-5}]$, the element 2 has no proper factor, so it is irreducible. But it is not prime because, although it divides the product $6 = (1 + \sqrt{-5})(1 - \sqrt{-5})$, it does not divide either factor.*

REMARK 6.16. *Since irreducible elements in a UFD are prime, the phrase* ***irreducible factorization*** *and* ***prime factorization*** *are synonymous. We can use them interchangeably when we are working in a UFD, but not otherwise.*

There is a simple way of deciding whether an element a divides another element b in a *UFD*, in terms of their irreducible (or prime) factorizations.

PROPOSITION 6.4. *Let R be a UFD, and let $a = p_1 \cdots p_r$, and $b = q_1 \cdots q_s$ be given prime factorization of two elements of R. Then a divides b if and only if $s \geq r$ and with a suitable ordering of the factors q_i of b, p_i is an associate of q_i for each $i = 1, ..., r$.* □

COROLLARY 6.4. *Let R be a UFD, and let a, b be elements of R which are not both zero. Then there exists a* ***greatest common divisor*** *$d = \gcd(a, b)$ of a, b, i.e., an element d in R satisfying the following properties:*

(1) *d divides both a and b.*
(2) *if an element f of R divides both a and b, then f divides d.* □

It follows immediately from the second condition that any two greatest common divisors of a and b are in fact associates.

REMARK 6.17. *However,* the greatest common divisor need not have the form $ar + bs$ *for some $r, s \in R$, unless we are working in a PID R.*

EXAMPLE 6.11. *In the next section, we will prove that $\mathbb{Z}[x]$ is a UFD. In this ring, 2 and x are irreducible elements, so that they are primes, and* $\gcd(2, x) = 1$. *But 1 is not a linear combination of these two elements with integer polynomial coefficients.*

We recall that every ideal in the ring $\mathbb{Z}$ of integers is principal, which is a very important property of $\mathbb{Z}$. We have also proved that in an integral domain, every prime element is irreducible (Lemma 6.3). In our approach of proving that a *PID* is a *UFD*, we prove the following proposition first:

PROPOSITION 6.5. *In a PID, every irreducible element is prime.*

PROOF. Let a be irreducible. Assume that $a|bc$ but a does not divide b. We consider the ideal (a, b) in R generated by a and b. Since we are working in a *PID*, this ideal is generated by a single element, say d of R, $(a, b) = (d)$. We also note that this ideal properly contains the ideal (a) since a does not divide b by hypothesis; hence d is not an associate of a. Now, $(a, b) = (d)$ also implies that $d|a$ and $d|b$. Then since a is irreducible, d, which is a non-associate of a, must be a unit. Thus, $(a, b) = (1)$, i.e., $1 = ar + bs$ for some r, s in the ring. Thus, $c = car + cbs$ is divisible by a, which is what we wanted to prove. □

THEOREM 6.5. *A Principal Ideal Domain (PID) is a UFD.*

PROOF. Let R be a *PID*. Then every irreducible element of R is prime by the above proposition. Thus, according to Proposition 6.3, we need only to prove the existence of factorizations for R. By Proposition 6.2, this is equivalent to showing that R contains no infinite increasing chain of principal ideals. We argue by contradiction. Suppose that $(a_1) \subset (a_2) \subset (a_3) \subset \dots$ is such a chain. Then since the union I of these ideals is an ideal of R, it is generated by a single element, say b, $I = (b)$. Then, b is contained in one of the (a_n); hence $I \subset (a_N) \subset (a_{N+1}) \subset \dots \subset I$ for some N. In other words, our ascending chain of principal ideals is stationary, which contradicts the assumption that $(a_n) \subset (a_{n+1})$ for each n, and this completes the proof. □

REMARK 6.18. *The converse of Theorem 6.5 is not true in general. The ring* $\mathbb{Z}[x]$ *of integer polynomials is a UFD, but it is not a PID.*

PROPOSITION 6.6. *Let R be a PID and let p be a non-zero element of R. Then the followings hold and moreover they are equivalent to each other:*

- **a:** *$R/(p)$ is a field if and only if p is irreducible.*
- **b:** *The maximal ideals in a PID are those principal ideals generated by irreducible elements.*

PROOF. Since an ideal M is maximal if and only if the quotient ring R/M is a field, the two parts (a) and (b) are equivalent. More precisely, the $(b) \Longrightarrow (a)$ direction is rather obvious, because $R/(p)$ is a field if and only if (p) is a maximal ideal, which is in turn equivalent by part (b) to the statement that p is irreducible. Now, for another direction $(a) \Longrightarrow (b)$, we suppose part (a). For an irreducible element p of R, $R/(p)$ is a field; hence (p) is a maximal ideal. Now let M be a maximal ideal of R. Since we are in a *PID*, it is generated by a single element, say p of R, $M = (p)$. Since $R/(p)$ is a field, p is irreducible by part (a).

Now, we are going to prove the equivalence in part (b). First, let M be a maximal ideal. Then since we are working in a *PID*, $M = (p)$ for some $p \in R$. If p has a proper divisor b, then $(p) \subset (b) \subset (1)$ properly, so that $M = (p)$ is not a maximal ideal; hence p must be irreducible. Secondly, for an irreducible p in R, we know that the only divisors of p are the units and the associates of p. Therefore, the only principal ideals containing (p) are (1) and (p), that is, there are no other proper ideals properly containing (p); hence (p) is a maximal ideal. □

Exercises for Chapter 6.3

EXERCISE 6.6. *Prove that if R is an integral domain, then two elements a, b are associates if and only if they differ by a unit factor.*

EXERCISE 6.7. *Prove that $(a, b)[a, b] = ab$ in a principal ideal domain.*

EXERCISE 6.8. *Prove that $\mathbb{Z}[i]$ is a PID. [Hint: It is a Euclidean domain.]*

EXERCISE 6.9. *Prove that the following rings are Euclidean domains:*

(1) $\mathbb{Z}[\omega]$, *where* $\omega = e^{2\pi i/3}$.
(2) $\mathbb{Z}[\sqrt{-2}]$.

4. Gauss's Lemma

We first note that Theorem 6.4 allows us to fairly freely manipulate polynomials in one variable over a field k. This is mainly because $k[x]$ is a *PID*. In most of this section we fix our bottom field k to be the field $\mathbb{Q}$ of rational numbers, until we begin to mention that $\mathbb{Q}$ is the field of fractions of a *UFD* $\mathbb{Z}$.

Among rational polynomials in one variable are integral polynomials, and the set $\mathbb{Z}[x]$ of all integral polynomials in one variable, whose elements are main objects of our study in this section, is certainly a subring of $\mathbb{Q}[x]$; hence it is also an integral domain.

In this section, we are going to prove that $\mathbb{Z}[x]$ is a *UFD*, and its proof definitely needs the so-called Gauss's Lemma for primitive polynomials in $\mathbb{Z}[x]$.

EXAMPLE 6.12. *We note that $\mathbb{Z}[x]$ is not a PID, because the ideal $(2, x)$ in $\mathbb{Z}[x]$ generated by the prime number 2, which is necessarily an irreducible element of $\mathbb{Z}[x]$, and the irreducible polynomial x, which is obviously an irreducible element of $\mathbb{Z}[x]$, is never generated by a single element in $\mathbb{Z}[x]$.*

Then the theory developed to prove that $Z[x]$ is a *UFD* easily leads us to establish the following important result: If R is a *UFD*, then so is $R[x]$.

Here is our beginning example of an irreducible factorization in $\mathbb{Z}[x]$:

EXAMPLE 6.13. $6x^3 + 9x^2 + 9x + 3 = 3(2x + 1)(x^2 + x + 1) \in \mathbb{Z}[x]$.

REMARK 6.19. *We do not mention such a factorization as a "prime" factorization until we prove that* $\mathbb{Z}[x]$ *is a UFD.*

As we see from this example, irreducible factorizations are slightly more complicated in $\mathbb{Z}[x]$ than in $\mathbb{Q}[x]$. First, the prime integers are necessarily irreducible elements of $\mathbb{Z}[x]$, so they may appear in an irreducible factorization of an integral polynomial, i.e., a polynomial in $\mathbb{Z}[x]$. But in $\mathbb{Q}[x]$ irreducible elements are irreducible polynomials, and vice versa. Second, the irreducible factor $2x + 1$ is not a monic polynomial. If we want to stay with integer coefficients, we can not ask for monic factors.

DEFINITION 6.5. *An integral polynomial* $f(x)$ *is called* ***primitive*** *if its coefficients* $a_0, ..., a_n$ *have no common integer factor except for the units* ± 1 *and if its highest coefficient* a_n *is positive.*

Thus, irreducibility question of integral polynomials $f(x)$ in one variable is basically reduced to irreducibility question of its associated primitive polynomials, because the "whole integer factor" of $f(x)$ is easily determined by computing the greatest common divisor of coefficients of $f(x)$. At any rate, the integer factors of a polynomial $f(x) = a_n x^n + \cdots + a_0$ in $\mathbb{Z}[x]$ are common divisors of its coefficients $a_0, ..., a_n$ of $f(x)$.

LEMMA 6.4. *Every non-constant irreducible integral polynomial with positive leading coefficient is necessarily primitive, but not conversely.* □

REMARK 6.20. *The converse of this lemma requires some additional condition, as is described in Proposition 6.8 below.*

LEMMA 6.5.

(1) *Every non-zero rational polynomial* $f(x) \in \mathbb{Q}[x]$ *can be written as a product*

$$f(x) = cf_0(x),$$

where c *is a rational number and* $f_0(x)$ *is a primitive integral polynomial in* $\mathbb{Z}[x]$.

(2) *Moreover, this expression for* f *is unique.*

(3) *The polynomial* $f(x)$ *has integer coefficients if and only if* c *is an integer.*

(4) *If (3) holds, then* $|c|$ *is the greatest common divisor of the coefficients of* $f(x)$*, and the sign of* c *is the sign of the leading coefficient of* $f(x)$.

DEFINITION 6.6. *The* rational number c *which appears in this lemma is called the* ***content*** *of* $f(x)$.

Thus, we observe that if $f(x)$ has integer coefficients, then the content divides $f(x)$ in $\mathbb{Z}[x]$. We also see that $f(x)$ is primitive if and only if its content is 1.

PROOF. To find $f_0(x)$, we first multiply $f(x)$ by an integer to clear the denominators in its coefficients. This will give us a polynomial $f_1(x)$ with integer coefficients. Then we factor out the greatest common divisor of the coefficients of $f_1(x)$ and adjust the sign of the leading coefficient. The resulting polynomial $f_0(x)$ is primitive, and $f = cf_0$ for some rational number c. This proves existence.

To prove uniqueness, suppose that $cf_0(x) = dg_0(x)$, where $c, d \in \mathbb{Q}$ and $f_0(x)$ and $g_0(x)$ are primitive polynomials. We will show that $c = d$ and $f_0(x) = g_0(x)$. Clearing denominators of both c and d reduces us to the case that c and d are integers. Let $\{a_i\}, \{b_i\}$ denote the coefficients of $f_0(x)$ and $g_0(x)$, respectively. Then $ca_i = db_i$ for each i. Since the greatest common divisor of $\{a_0, ..., a_n\}$ is 1, c is the greatest common divisor of $\{ca_0, ..., ca_n\}$. Similarly, d is a greatest common divisor of $\{db_0, ..., db_n\} = \{ca_0, ..., ca_n\}$. Thus, $c = \pm d$ and $f_0(x) = \pm g_0(x)$. Then since the primitive polynomials $f_0(x)$ and $g_0(x)$ have positive leading coefficients, we conclude that $f_0(x) = g_0(x)$, and $c = d$.

Now, if $f(x)$ has integer coefficients, clearing of the denominators is not necessary; hence c is an integer, and up to sign it is the greatest common divisor of the coefficients, as stated. □

Now, we are ready to prove Gauss's Lemma:

THEOREM 6.6. *(Gauss's Lemma) A product of two primitive polynomials in $\mathbb{Z}[x]$ is again primitive.*

PROOF. Let the primitive polynomials be denoted by f and g, and let $h = fg$ be their product. Since the leading coefficients of f and g are positive, the leading coefficient of h is also positive. To prove that h is primitive, that is, to prove that no prime integer p divides all the coefficients of h, we consider the ring homomorphism $\mathbb{Z}[x] \longrightarrow \mathbb{Z}_p[x]$ for a prime integer p, which is introduced in Remark 6.9. Since both f and g are primitive, no prime integer p will divide their integer coefficients, and hence both residues $\bar{f}$ and $\bar{g}$ are not the zero polynomial in $\mathbb{Z}_p[x]$ for each prime integer p. Then since $\mathbb{Z}_p[x]$ is an integral domain, their product $\bar{f}\bar{g}$, which is exactly the same as $\bar{h} = \overline{fg}$, is also a non-zero polynomial in $\mathbb{Z}_p[x]$ for each prime integer p. This exactly means that $h \notin p\mathbb{Z}[x]$, the kernel of the ring homomorphism, for each prime integer p. Thus, h is primitive. □

REMARK 6.21. *The following proposition relates divisibility between two integral polynomials $f(x)$ and $g(x)$ in $\mathbb{Z}[x]$ with divisibility between the same $f(x)$ and $g(x)$ in $\mathbb{Q}[x]$.*

More precisely, we note that a division in an integral domain is always a division in its field of fractions, so our movement in this direction gives no difficulty to us. On the other hand, if coefficients of polynomials are all invertible, i.e., units, then it is quite easy to handle coefficients of polynomials. Thus, we have a tendency of preferring to work on divisions over a field. But, in order to properly relate a division over a field of fractions of a UFD R to a division in its subring R, where our objects for *division are assumed to be in the subring R, we definitely need Gauss's Lemma, as the proof of the following Proposition demonstrates.*

The main reason why we prefer to work on a UFD is to keep a good track on each irreducible factor of the involved objects while we are in division process.

We also recall that every field has only two ideals, (0) *and* (1), *so it is a PID; hence it is a UFD.*

PROPOSITION 6.7.

- **a:** *Let f, g be rational polynomials in $\mathbb{Q}[x]$, and let f_0, g_0 be their associated primitive polynomials in $\mathbb{Z}[x]$. If f divides g in $\mathbb{Q}[x]$, then f_0 divides g_0 in $\mathbb{Z}[x]$.*
- **b:** *Let f be a primitive polynomial in $\mathbb{Z}[x]$, and let g be any integral polynomial in $\mathbb{Z}[x]$. Suppose that f divides g in $\mathbb{Q}[x]$, say $g = fq$, with a rational polynomial q in $\mathbb{Q}[x]$. Then q is an integral polynomial in $\mathbb{Z}[x]$, and hence f divides g in $\mathbb{Z}[x]$.*
- **c:** *Let f, g be integral polynomials in $\mathbb{Z}[x]$. If they have a common non-constant rational polynomial factor in $\mathbb{Q}[x]$, then they have a common non-constant integral polynomial factor in $\mathbb{Z}[x]$ too, and conversely.*

PROOF. The converse of each statement is trivial.

To prove part a, we may clear denominators so that f and g become primitive. Then part a is a consequence of part b. More precisely, according to Lemma 6.5 we write $f = cf_0$, $g = dg_0$, where c and d are associated contents in $\mathbb{Q}$, respectively, and f_0, g_0 are associated primitives, respectively. Thus, if f divides g in $\mathbb{Q}[x]$, then we have $dg_0 = \alpha c f_0$, with some $\alpha \in \mathbb{Q}[x]$, so that $g_0 = d^{-1}\alpha c f_0$. Thus, the primitive f_0 divides the integral polynomial g_0 in $\mathbb{Q}[x]$. Thus, part a is a consequence of part b.

To prove part b, we apply Lemma 6.5 in order to write the quotient in the form $q = cq_0$, where q_0 is primitive and $c \in \mathbb{Q}$. Then by Gauss's Lemma (Theorem 6.6), fq_0 is primitive, and the equation $g = cfq_0$ shows that fq_0 is the primitive polynomial g_0 associated to g. Therefore $g = cg_0$

is the unique expression for g referred to in Lemma 6.5, and the rational number c is the content of g. Then since g is an integral polynomial by hypothesis, it follows that c is in fact an integer, and hence q is also an integral polynomial.

Finally, to prove part c, suppose that f, g have a common rational polynomial factor h in $\mathbb{Q}[x]$. We may assume that h is primitive, and then by part b, h divides both f and g in $\mathbb{Z}[x]$. □

COROLLARY 6.5. *If a non-constant integral polynomial f is irreducible in $\mathbb{Z}[x]$, then it is also irreducible in $\mathbb{Q}[x]$.*

PROOF. This is true by part b of the proposition above. □

PROPOSITION 6.8. *Let f be an integral polynomial with positive leading coefficient. Then f is irreducible in $\mathbb{Z}[x]$ if and only if either*

(1) *f is a prime integer, or*
(2) *f is a primitive polynomial which is irreducible in $\mathbb{Q}[x]$.*

PROOF. Suppose that f is irreducible. As in Lemma 6.5, we may write $f = cf_0$, where f_0 is primitive. Since f irreducible, this can not be a proper factorization; hence either c or f_0 is 1. If $f_0 = 1$, then f is constant, and to be irreducible, a constant polynomial must be a prime integer. Otherwise, i.e., if $c = 1$, then f is primitive, and it is irreducible in $\mathbb{Q}[x]$ by Corollary 6.5.

The converse, that integer primes and primitive irreducible polynomials are irreducible element of $\mathbb{Z}[x]$, is clear. □

EXAMPLE 6.14. *The polynomial $3(2x+1)$ is irreducible over $\mathbb{Q}$ but is not irreducible over $\mathbb{Z}$.*

Let $f(x)$ be an integral polynomial in $\mathbb{Z}[x]$. Then it is also in $\mathbb{Q}[x]$. Then since $\mathbb{Q}[x]$ is a *PID*, it is a *UFD*; hence we have a unique factorization of $f(x)$ in $\mathbb{Q}[x]$ up to a unit:

$$f(x) = cq_1(x) \cdots q_n(x),$$

where c is a unit and each $q_i(x)$ are rational irreducible monic polynomials in $\mathbb{Q}[x]$.

In order to derive a factorization of $f(x)$ in $\mathbb{Z}[x]$ from this factorization, we simply apply the following procedure: First, we clear out the denominator of $q_1(x)$ and take out the greatest common divisor of all integer coefficients of the resulting integral polynomial in order to have in $\mathbb{Q}[x]$

$$f(x) = c_1 q_1'(x) q_2(x) \cdots q_n(x),$$

where c_1 is another unit in $\mathbb{Q}$, $q_1'(x)$ is primitive polynomial in $\mathbb{Z}[x]$ which is necessarily irreducible there, and $q_2(x) \cdots q_n(x)$ is still a rational polynomial in $\mathbb{Q}[x]$. This says that the integral polynomial $f(x)$ is divided by a primitive polynomial $q_1'(x)$ in $\mathbb{Q}[x]$. Therefore, by part b of Proposition 6.7 the product $c_1 \cdot q_2(x) \cdots q_n(x)$ is in fact an integral polynomial in $\mathbb{Z}[x]$. But since this is not yet good enough, we then apply the same treatment to $q_2(x)$ as we have done to $q_1(x)$ to have

$$f(x) = c_2 q_1'(x) q_2'(x) q_3(x) \cdots q_n(x),$$

where c_2 is another unit in $\mathbb{Q}$, $q_2'(x)$ is primitive polynomial in $\mathbb{Z}[x]$ which is necessarily irreducible there, and $q_3(x) \cdots q_n(x)$ is a rational polynomial in $\mathbb{Q}[x]$. If we repeat this "conversion" process at most n times, we will end up with

$$f(x) = c_n q_1'(x) \cdots q_n'(x),$$

where c_n is a unit in $\mathbb{Q}$, each $q_i'(x)$ primitive irreducible polynomial in $\mathbb{Z}[x]$; hence the product $q_1'(x) \cdots q_n'(x)$ is again primitive by Gauss's Lemma (Theorem 6.6). Then, again by part b of Proposition 6.7 whose proof also requires Gauss's Lemma, c_n is in fact an integer! In this way, we have derived a factorization of $f(x)$ in $\mathbb{Z}[x]$ from its always-existing unique factorization in the *UFD* $\mathbb{Q}[x]$. *In other words, we have proved that existence of factorization holds in* $\mathbb{Z}[x]$.

Having this process at our hands, we are on our way in proving that $\mathbb{Z}[x]$ is a *UFD* using Proposition 6.3.

PROPOSITION 6.9. *Every irreducible element of* $\mathbb{Z}[x]$ *is a prime element.*

PROOF. Let f be irreducible, and suppose that f divides in $\mathbb{Z}[x]$ a product gh, where $g, h \in \mathbb{Z}[x]$. By Proposition 6.8 we have only two cases.

Case 1. $f = p$ is a prime integer. We write $g = cg_0$ and $h = dh_0$ with integers c, d, as in Lemma 6.5. Then g_0h_0 is again primitive by Gauss's Lemma, and hence some coefficient a of the expansion of the product g_0h_0 must not be divisible by p. On the other hand, since p divides the product gh, p must divides all the coefficients of the expansion of the product gh; hence in particular p must divide the integer product acd. Then since p does not divide a, it must divide either c or d; hence p divides either g or h.

Case 2. f is a primitive polynomial which is irreducible in $\mathbb{Q}[x]$. In this case, f is a prime element of $\mathbb{Q}[x]$ since $\mathbb{Q}[x]$ is a *PID*. Thus, f divides either g or h in $\mathbb{Q}[x]$. Then by part b of Proposition 6.7, f divides either g or h in $\mathbb{Z}[x]$. □

THEOREM 6.7.

(1) *The polynomial ring* $\mathbb{Z}[x]$ *is a Unique Factorization Domain (UFD).*
(2) *Every non-zero integral polynomial* $f(x)$ *in* $\mathbb{Z}[x]$ *which is not* ± 1 *can be written as a product*

$$f(x) = \pm p_1 \cdots p_m \cdot q_1(x) \cdots q_n(x),$$

where the p_i *are prime integers and the* $q_i(x)$ *are irreducible primitive polynomials in* $\mathbb{Z}[x]$.
(3) *The expression is unique up to arrangement of the factors.*

PROOF. We have already seen existence of factorizations in $\mathbb{Z}[x]$ above. Thus, by Proposition 6.9 and Proposition 6.3 we completes our proof. □

Now let R be a *UFD*, and let F be its field of fractions. Then $R[x]$ is a subring of $F[x]$, and the results of this section can be copied, replacing $\mathbb{Z}$ with R and $\mathbb{Q}$ with F throughout. The only change to be made is that instead of normalizing primitive polynomials it is better to allow ambiguity caused by unit factors, as in the previous section. The main results are these:

THEOREM 6.8. *Let R be a UFD with field of fractions F.*

- **a:** *Let f, g be polynomials in $F[x]$, and let f_0, g_0 be their associated primitive polynomials in $R[x]$. If f divides g in $F[x]$, then f_0 divides g_0 in $R[x]$.*
- **b:** *Let f be a primitive polynomial in $R[x]$, and let g be any polynomial in $R[x]$. Suppose that f divides g in $F[x]$, say $g = fq$, with a polynomial q in $F[x]$. Then q is a polynomial in $R[x]$, and hence f divides g in $R[x]$.*
- **c:** *Let f, g be polynomials in $R[x]$. If they have a common non-constant factor in $F[x]$, then they have a common non-constant factor in $R[x]$ too, and conversely.*
- **d:** *If a non-constant polynomial f is irreducible in $R[x]$, then it is irreducible in $F[x]$, and conversely.*
- **e:** *$R[x]$ is a UFD.*

The proof of Theorem 6.8 follows the pattern established for the ring $\mathbb{Z}[x]$, and we omit it. □

Since $R[x_1, ..., x_n]$ is isomorphic to $R[x_1, ..., x_{n-1}][x_n]$, we obtain the next corollary:

COROLLARY 6.6. *The polynomial rings $\mathbb{Z}[x_1, ..., x_n]$ and $F[x_1, ..., x_n]$, where F is a field, are UFD's.* □

Thus, the ring $\mathbb{C}[x, y]$ of complex polynomials in two variables is a UFD.

REMARK 6.22. *In contrast to the case of one variable, however, where every complex polynomial is a product of linear ones by Gauss' Fundamental Theorem of Algebra, Theorem 1.1, complex polynomials in two variables are often irreducible, and hence prime.*

Exercises for Chapter 6.4

EXERCISE 6.10. *Let a, b be elements of a field F, with $a \neq 0$. Prove that a polynomial $f(x)$ over F is irreducible over F if and only if $f(ax+b)$ is irreducible over F.*

EXERCISE 6.11. *Let $F = \mathbb{C}(x)$, and let $f(x,y), g(x,y) \in \mathbb{C}[x,y]$. Prove that if $f(x,y)$ and $g(x,y)$ have a common factor in $F[y]$, then they also have a common vector in $\mathbb{C}[x,y]$.*

EXERCISE 6.12. *Prove that two integer polynomials are relatively prime in $\mathbb{Q}[x]$ if and only if the ideal they generate in $\mathbb{Z}[x]$ contains an integer.*

5. Explicit Factorization of Polynomials

We first recall that over the field $\mathbb{C}$ of complex numbers, thanks to Gauss' Fundamental Theorem of Algebra, every polynomial of positive degree has a complex root α and hence has a divisor of the form $x - \alpha$. Therefore, the irreducible polynomials over $\mathbb{C}$ are linear, and the irreducible factorization of a polynomial over $\mathbb{C}$ has the form

$$g(x) = c(x - \alpha_1) \cdots (x - \alpha_n), \tag{6.4}$$

where α_i are the roots of $g(x)$, repeated as necessary. Thus, the uniqueness of this factorization is not surprising.

And, over the field $\mathbb{R}$ of real numbers, there are only two classes of irreducible polynomials: linear polynomials and irreducible quadratic polynomials. We recall that a real quadratic polynomial $x^2 + bx + c$ is irreducible over $\mathbb{R}$ if and only if its discriminant $b^2 - 4c$ is negative, in which case it has a pair of complex conjugate roots.

What we want are the irreducible factors of a rational polynomial $g(x)$ in $\mathbb{Q}[x]$. By clearing out the denominator of $g(x)$ and pulling out the gcd of the resulting polynomial, we can write

$$g(x) = cg_0(x),$$

where $g_0(x)$ is a primitive integral polynomial in $\mathbb{Z}[x]$, which is what Lemma 6.5 all about, and hence by Corollary 6.5 this amounts to determining the irreducible factors in $\mathbb{Z}[x]$. Thus, in this section we want to seek explicit methods of determining the factors of a given integral polynomial

$$f(x) = a_n x^n + \cdots + a_1 x + a_0. \tag{6.5}$$

Linear factors of $f(x)$ can be found fairly easily: If $b_1 x + b_0$ divides $f(x)$, then b_1 should divide a_n and b_0 should divide a_0. Then since there are only finitely many integers which divide a_n and a_0, we can try all possibilities. In each case, we carry out the division and determine whether the remainder is zero or not. Or, we may substitute the rational number $r = -b_0/b_1$ into $f(x)$ to see if it is a root.

REMARK 6.23. *Though things are not so clear for factors of higher degree, it is known that Kronecker (1823 - 1891) showed that the factors can be determined with a finite number of computations. His method is based on the Lagrange interpolations formula, but unfortunately this method requires too many steps to be practical except for factors of low degree, and a lot of work has been done on the problem of efficient computations.*

We we have mentioned before, one of the most useful methods is then computation modulo p, using the homomorphisms $\mathbb{Z}[x] \longrightarrow \mathbb{Z}_p[x]$ defined by

$$f(x) = a_n x^n + \cdots + a_0 \longmapsto \bar{f}(x) = \bar{a}_n x^n + \cdots + \cdot\bar{a}_0,$$

for various prime integers p, where $\mathbb{Z}_p := \mathbb{Z}/p\mathbb{Z}$ is a field for each prime number p. If our polynomial $f(x)$ factors in $\mathbb{Z}[x]$, so is, its residue $\bmod\, p$ under this map for each prime integer p, i.e., its image in $\mathbb{Z}_p[x]$ for each prime integer p. Then since there are only finitely many polynomials of the same degree in $\mathbb{Z}_p[x]$, all factorizations there can be carried out in finitely many steps.

PROPOSITION 6.10. *Let $f(x) = a_n x^n + \cdots + a_1 x + a_0 \in \mathbb{Z}[x]$ be an integral polynomial, and let p be a prime integer which does not divide* *the leading coefficient* a_n. *If the residue* $\bar{f}(x)$ *of* $f(x) \bmod p$ *in* $\mathbb{Z}_p[x]$ *is irreducible there, then* $f(x)$ *is irreducible in* $\mathbb{Q}[x]$, *but not conversely.*

PROOF. This follows from an inspection of the homomorphism $\mathbb{Z}[x] \longrightarrow \mathbb{Z}_p[x]$. We need the assumption that p does not divide a_n in order to rule out the possibility that a factor $g(x)$ of $f(x)$ could reduce to a constant in $\mathbb{Z}_p[x]$. This assumption is preserved if we replace $f(x)$ with its associated primitive polynomial. Thus we may assume that $f(x)$ is primitive. Since p does not divide the leading coefficient a_n, the degrees of $f(x)$ and $\bar{f}(x)$ are the same. Now, if $f(x)$ factors in $\mathbb{Q}[x]$, by Corollary 6.5 it also factors in $\mathbb{Z}[x]$. Let $f(x) = g(x)h(x)$ be a proper factorization in $\mathbb{Z}[x]$. Since $f(x)$ is primitive, $g(x)$ and $h(x)$ have positive degree. Since $\deg(f(x)) = \deg(\bar{f}(x))$ and $\bar{f}(x) = \bar{g}(x)\bar{h}(x)$, it follows that $\deg(g(x)) = \deg(\bar{g}(x))$ and $\deg(h(x)) = \deg(\bar{h}(x))$; hence $\bar{f}(x) = \bar{g}(x)\bar{h}(x)$ is a proper factorization in $\mathbb{Z}_p[x]$, which shows that $\bar{f}(x)$ is reducible there. Then by Corollary 6.5 again, we complete our proof. □

The converse of this proposition does not hold in general.

EXAMPLE 6.15. *One easily checks by trial and error that the polynomial* $f(x) = x^4 - 10x^2 + 1$ *is irreducible over* $\mathbb{Z}$. *But,* $\bar{f}(x) \bmod p$ *for* $p = 2, 3, 5, 7$ *at least, is reducible over* $\mathbb{Z}_p$:

$$\begin{aligned} f(x) \bmod 2 &= x^4 + 1 = (x^2 + 1)^2, \\ f(x) \bmod 3 &= x^4 + 2x^2 + 1 = (x^2 + 1)^2, \\ f(x) \bmod 5 &= x^4 + 1 = (x^2 + 2)(x^2 + 3), \\ f(x) \bmod 7 &= x^4 + 4x^2 + 1 = (x^2 + 6x + 6)(x^2 + x + 6). \end{aligned}$$

Can you show that $\bar{f}(x) \bmod p$ *is in fact irreducible over* $\mathbb{Z}_p$ *for some prime* p?

Thus, the method of reduction modulo p will not always work. But it does work quite often:

Suppose that we suspect that a given integral polynomial $f(x) \in \mathbb{Z}[x]$ is irreducible. Then we can try reduction modulo p for a few low primes,

$p = 2$ or 3 for instance, and hope that $\bar{f}(x)$ turns out to be of the same degree and irreducible. If so, we will have proved that $f(x)$ is indeed irreducible over $\mathbb{Z}$. Since $\mathbb{Z}_p$ is a field, $\mathbb{Z}_p[x]$ is a *PID*; hence Theorem 6.4 still holds for $\mathbb{Z}_p[x]$.

Before we go further, we want to explain a method, called the **sieve of Eratosthenes**, of determining the prime integers p less than a given number n. The method goes as follows: We list the numbers from 2 to n. We know that 2 is prime, then we cross out all multiples of 2 from our list. Then 3 is prime, since it is not divisible by any smaller prime. Then we cross out all multiples of 3 from our list, and we keep going on until we reach up to n. Then the left-overs are prime numbers less than n.

This "**sieve method**" will also determine the irreducible polynomials in $\mathbb{Z}_p[x]$. We list all polynomials, degree by degree, and the cross out products.

EXAMPLE 6.16. *The linear polynomials in $\mathbb{Z}_2[x]$ are x and $x+1$, and there are irreducible there. The polynomials of degree 2 are x^2, x^2+1, x^2+x, and x^2+x+1. Then the first three are divisible by x or $x+1$, so the last one is the only irreducible polynomial of degree 2 in $\mathbb{Z}_2[x]$. We keep going on for polynomials of degree 3, and so on. Here is a list of irreducible polynomials of degree ≤ 4 over $\mathbb{Z}_2$:*

$$
\begin{aligned}
&x, x+1;\\
&x^2+x+1;\\
&x^3+x^2+1, x^3+x+1;\\
&x^4+x^3+1, x^4+x+1, x^4+x^3+x^2+x+1.
\end{aligned}
$$

By trying the polynomials in the above list, we can factor all polynomials of degree $9 = (4+3+2+1)-1$ or less in $\mathbb{Z}_2[x]$, and hence we can extend this list to another list of irreducible polynomials of degree ≤ 9 in $\mathbb{Z}_2[x]$, etc. and etc.

EXAMPLE 6.17. *Thus, $x^4-6x^3+12x^2-3x+9$ is irreducible in $\mathbb{Q}[x]$; and hence in $\mathbb{Z}[x]$ because it is primitive. For, its residue in $\mathbb{Z}_2[x]$ is x^4+x+1*

and it is irreducible there. We note that we can not conclude that it is irreducible over $\mathbb{Q}$ *from Eisenstein's Criterion in the next Proposition, because the constant term is divisible by* 3^2.

EXAMPLE 6.18. *The monic irreducible polynomials over degree* 2 *over* $\mathbb{Z}_3$ *are the followings:*

$$x^2 + 1, x^2 + x - 1, x^2 - x - 1.$$

We also note that reduction modulo p may help describe the factorization of a polynomial even though the residue is reducible:

EXAMPLE 6.19. *Consider the polynomial* $f(x) = x^3 + 6x + 3$ *for example. Reducing modulo* 3, *we obtain* x^3 *and this does not look like quite promising. However, suppose that* $f(x)$ *were reducible, say* $(ax+b)(cx^2 + dx + e) = x^3 + 6x + 3$. *Then the residue of* $ax + b$ *would have to divide* x^3 *in* $\mathbb{Z}_3[x]$, *which would imply* $b \equiv 0 \bmod 3$. *Similarly, we could conclude* $e \equiv 0 \bmod 3$. *But it is impossible to satisfy both of these conditions, because* $be = 3$. *Therefore, no such factorization exists, and* $f(x)$ *is irreducible over* $\mathbb{Q}$.

The principle at work in this Example is called the **Eisenstein Criterion**.

PROPOSITION 6.11. *(Eisenstein Criterion) Let* $f(x) = a_n x^n + \cdots + a_0 \in \mathbb{Z}[x]$ *be an integral polynomial, and let* p *be a prime integer. Suppose that the coefficients of* $f(x)$ *satisfies the following conditions:*

(1) *p does not divide the leading coefficient* a_n;
(2) *p divides all other coefficients* $a_{n-1}, \ldots, a_0$;
(3) *p^2 does not divide the constant term* a_0.

Then f is irreducible in $\mathbb{Q}[x]$. *If moreover f is primitive, then it is irreducible in* $\mathbb{Z}[x]$.

PROOF. Let $\bar{f}(x)$ denote the reside $\bmod\, p$ of $f(x)$. The first two hypotheses imply that $\bar{f}(x) = \bar{a}_n x^n$ and that $\bar{a}_n \neq 0 \bmod p$. If $f(x)$ were reducible in $\mathbb{Q}[x]$, then it would factor in $\mathbb{Z}[x]$ into factors of positive degree, say $f(x) = g(x)h(x)$ by Corollary 6.5. Then $\bar{g}(x)$ and $\bar{h}(x)$ would divide $\bar{a}_n x^n$, and hence each of these polynomials would be a monomial. Therefore all coefficients of $g(x)$ and $h(x)$, except the leading terms, would be divisible by p. But, let the constant coefficients of $g(x), h(x)$ be b_0, c_0. Then the constant coefficient of $f(x)$ is obviously $a_0 = b_0 c_0$. Since p divides both b_0 and c_0, p^2 divides a_0, which contradicts the third hypothesis. Therefore, $f(x)$ must be irreducible in $\mathbb{Q}[x]$. The last assertion follows from Proposition 6.8. □

One of the most important applications of the Eisenstein Criterion is to prove the irreducibility of the cyclotomic polynomial

$$x^{p-1} + x^{p-2} + \cdots + x + 1$$

is irreducible in $\mathbb{Q}[x]$, whose roots are the p-th root of unity, the powers of $\varsigma = e^{2\pi i/p}$:

COROLLARY 6.7. *Let p be a prime. The polynomial $f(x) = x^{p-1} + x^{p-2} + \cdots x + 1$ is irreducible in $\mathbb{Q}[x]$.*

PROOF. We first know that $(x-1)f(x) = x^p - 1$. If we make a change of coordinate $x = y + 1$, then it become

$$\begin{aligned} yf(y+1) &= (y+1)^p - 1 \\ &= y^p + \binom{p}{1} y^{p-1} + \binom{p}{2} y^{p-2} + \cdots + \binom{p}{p-1} y, \end{aligned}$$

where $\binom{p}{i} = p(p-1)\cdots(p-i-1)/i! \in \mathbb{N}$ is the i-th binomial coefficient. Thus, by dividing out both side by y, we obtain

$$f(y+1) = y^{p-1} + \binom{p}{1} y^{p-2} + \binom{p}{2} y^{p-3} + \cdots + \binom{p}{p-1}.$$

Now, for any $i < p$, p never divides each $i - j$ for $j = 0, 1, ..., i$; hence p can not divide $i!$. Thus, $\binom{p}{i}$ is divisible by p for $i < p$. Therefore, by Eisenstein Criterion $f(y+1)$ is irreducible in $\mathbb{Q}[y]$. Thus, $f(x)$ must be irreducible over $\mathbb{Q}$. □

Exercises for Chapter 6.5

EXERCISE 6.13. *Show that the following polynomials are irreducible over* $\mathbb{Q}$.

(1) $x^2 + 27x + 213$.
(2) $8x^3 - 6x + 1$.
(3) $x^5 - 3x^4 + 3$.

EXERCISE 6.14. *Factor* $x^5 + 5x + 5$ *into irreducible factors in* $\mathbb{Q}[x]$ *and in* $\mathbb{F}_2[x]$.

EXERCISE 6.15. *Factor* $x^3 + x + 1$ *in* $\mathbb{F}_p[x]$, *when* $p = 2, 3, 5$.

EXERCISE 6.16. *Factor* $x^4 + x^2 + 1$ *into irreducible factors in* $\mathbb{Q}[x]$.

EXERCISE 6.17. *Let* p *be a prime integer. Prove that the polynomial* $x^n - p$ *is irreducible over* $\mathbb{Q}$.

EXERCISE 6.18. *Prove that the polynomial* $x^2 + y^2 - 1$ *is irreducible over* $\mathbb{Q}$ *by the method of undetermined coefficients, that is, by studying the equation*

$$(ax + by + c)(a'x + b'y + c') = x^2 + y^2 - 1,$$

where a, b, c, a', b', c' *are unknown.*

6. Additive Factorization Theory?

Before we close the current section, we try to introduce very briefly the subject, called, the additive number theory, in which we study, for instance, the number $p(n)$ of different way of adding smaller numbers which is equal to a given natural number n. For example, we have in total 7 different ways of expressing the number 5:

$$
\begin{aligned}
&5,\\
&4+1,\\
&3+2,\\
&3+1+1,\\
&2+2+1,\\
&2+1+1+1,\\
&1+1+1+1+1.
\end{aligned}
$$

Here, we consider, for instance, $4 + 1$ is the "same" as $1 + 4$. Thus, $p(5) = 7$. The function $p(n)$ is called the "partition function", and many interesting questions on the property of $p(n)$ are answered, basically starting with the works of Euler, and then MacMahon, Hardy, Ramanujan, and Rademacher, etc. For example, Euler proved

THEOREM 6.9. *(Euler) Given a natural number n, one has*

$$p(n) = p(n-1)+p(n-2)-p(n-5)-p(n-7)+\cdots+(-1)^{j+1}p(n-n_j), \tag{6.6}$$

where $n_j = j(3j \pm 1)/2$ are the so-called pentagon numbers.

It is well-known that P.A. MacMahon used this recurrence formula in 1916 to compute the values of $p(n)$ for n up to $n = 200$, which is obtained from the famous identity of Euler:

THEOREM 6.10. *(Euler's pentagonal number theorem)*

$$\prod_{j=1}^{\infty}(1-t^j) = 1 + \sum_{j=1}^{\infty}(-1)^j t^{j(3j-1)/2}(1+t^j). \tag{6.7}$$

The proof of these two theorems are nowadays a standard material in most books in Number Theory.

According to George Andrew's account in p. 150 of his book, The Theory of Partitions, G.H. Hardy and S. Ramanujan were the first mathematicians who gave an asymptotic formula for $p(n)$ by power series in

1917, the first term of which is

$$\frac{1}{2\pi\sqrt{2}}\frac{d}{dn}\left(\frac{\exp(\frac{2\pi}{\sqrt{6}}\sqrt{n-\frac{1}{24}}}{\sqrt{n-\frac{1}{24}}}\right).$$

The first five terms of the Hardy-Ramanujan formula give the "correct" value of $p(200)$, which is $3,972,999,029,388$.

In 1937, H. Rademacher discovered an exact formula for $p(n)$ by a convergent series, which yields Hardy and Ramanujan's asymptotic formula as a corollary.

THEOREM 6.11. *(Rademacher) Given a natural number n, one has*

$$p(n) = \frac{1}{\pi\sqrt{2}}\sum_{k=1}^{\infty} A_k(n)\sqrt{k}\frac{d}{dn}\sqrt{\frac{\sinh((\pi/k)(2/3)(n-\frac{1}{24}))}{n-\frac{1}{24}}}, \tag{6.8}$$

where

$$A_k(n) = \sum_{\substack{h \ mod \ k \\ (h,k)=1}} \varpi_{h,k}\exp(-2\pi i n h/k),$$

where

$$\varpi_{h,k} = \exp(\pi i s(h,k)),$$

is a $24k$-th root of unity and where

$$s(h,k) = \sum_{\mu=1}^{k-1}\left(\frac{\mu}{k} - \left[\frac{\mu}{k}\right] - \frac{1}{2}\right)\left(\frac{h\mu}{k} - \left[\frac{hk}{k}\right] - \frac{1}{2}\right)$$

is the so-called Dedekind sum.

It is well-known that this theorem is based on the theory of modular forms and the proof of this hard-to-understand theorem is well-presented, for instance, in pp. 71-81 of George Andrew's book, The Theory of Partitions. This is the formula mathematicians up to now insisted to call an exact formula for $p(n)$, although it is very analytic in nature.

On the other hand, since the size of $p(n)$ increases very rapidly, for instance,

$$\begin{aligned} p(10) &= 42, \\ p(100) &= 190569292, \\ p(1000) &= 24061467864032622473692149727991, \end{aligned}$$

it requires a good computational means, for example, CAS(Computer Algebra System) like MAPLE or REDUCE for "Symbolic Computation", to compute the values $p(n)$ for various relatively large n, even if these theoretical results are available to us.

Exercises for Chapter 6.6

EXERCISE 6.19. *Using Theorem 6.10, prove Theorem 6.9.*

APPENDIX A

An Introduction to Algebraic Analysis

1. Tools in Mathematics

The whole body of Mathematics consists of many different kinds of knowledge, or results and procedures which produce such results, most of which are verified through the process of the so-called "proof". Those proved and somewhat useful or important are usually called **theorems**. Unproven ones are formally called **conjectures**. The place for most mathematicians should be between these two words, because we are born to establish reasonable conjectures and then prove them. When one tries to prove something, something itself must be first recognized as a reasonable mathematical object; discovering such object could be quite personal.

In proving reasonable conjectures, one generally uses two tools, a special type of logics, called **mathematical logic**, and **computation,** definitely guided by his own mathematical intuitions or experiences. Since they are applied as tools in the process of the so-called **proof**, they themselves cannot be called mathematics. However, the kinds of the methods employed certainly put some unavoidable limitations on the character of the consequences. These two tools, nevertheless, do not hurt each other. In contrast, they help each other, and these two are the most powerful tools we, mathematicians, have. Depending on the degree of putting more emphasis on any of these two, one usually has two basic tendencies in doing mathematics, axiomatic or computational.

1.1. Axiomatic Nature. Although it is very difficult to figure out what constitutes mathematics, one may describe its axiomatic nature as follows: Once a reasonable mathematical conjecture is recognized, one accepts, without questioning their existence or validity, a finite number of undefined terminologies and that of postulates (i.e., axioms or principles) which describe some relations between the undefined terminologies, and one applies a finite number of steps of mathematical logic directly to them or to some of their consequences until one finally reaches the desired positive or negative conclusions. As a demonstrative science started even before Greek period, mathematics has shown this axiomatic nature for a long time. One can find such an evidence in, for instance, Euclid's Elements [**E**] and other works of Greek mathematicians and philosophers. In fact, one can even trace back to their works to see that most of the formal terminologies used in mathematics, something like definitions, propositions, corollaries, etc., were already used in his Elements and other Greek literatures. The meaning and usage of these terminologies and other metamathematical concerns are also well explained in the introduction of A. Whitehead and B. Russell's famous work, *Principia Mathematica* [**WR**], which was originally published in 1910.

1.2. Computational Nature. When computations in general are concerned, we know that there are basically two different kinds, namely, exact computation and approximation. **Approximation** is done through the limiting process mostly in terms of inequalities, where as **Exact Computation or Symbolic Computation** is done through the so-called four binary operations: addition, subtraction, multiplication and acceptable division. These operations are applied not only to numbers but also to a mixture of numbers and **symbols**.

Computations are very important to us at least for the following three reasons. First, they usually show up in the middle of the process called proof. This is almost unavoidable to every mathematician. Secondly, a decent amount of computational experiments quite often reveals to us what's going on in a very clean manner, while simple-minded 'what if ...' ways of logical questioning cannot. Thirdly, we, mathematicians, can communicate with other branches of sciences very efficiently and successfully, through computational results and their usefulness. Considering that mathematics has so far grown up to this much within the accumulation of our civilization, this point of view is also very important, because isolation only hurts oneself till one's starvation.

Up to the end of the 19th century or the beginning of the 20th century, some important mathematical works discovered and proved were based on individual mathematician's heavy hand-computational experiments. However, as was the case in Invariant Theory of the 19th century, mathematicians were practically forced out to give up heavy computations when the number of unknowns and the size of numbers are substantially increased so that hand-calculation obviously becomes terrible. Any attempt based on computational experiments may be called a **computational approach**.

2. Overemphasis on Axiomatic Approach in the 20th Century

The revival of the so-called **axiomatic approach** after the Greek period (namely, Euclid's Elements) is clearly evident in the Lecture Notes of E. Artin and E. Noether, written up by B.L. van der Waerden in 1930s [**BW**]. This approach, especially well-supported by precise notations adapted from the subject, Set Theory, was so good in pursuing a particular-minded concept or property, that it rapidly pushed out the old tradition of computational approaches, which was quite common up to the first quarter of 20th century, from the mathematical world. In addition, some eminent mathematicians, like Banach in Analysis or later E. Spanier in Algebraic Topology [**ES**], tried to figure out and establish sufficiently small number of axioms but still good enough to explain successfully the rest of the theory they know. In order to fulfill such tasks, they examined many computed examples of different kinds and pulled out some common underlying axioms which replace too much involved terrible computations. The success of this attempt is quite evident in almost every part of current mathematics in both research articles and textbooks. It is also around this period that most mathematicians officially but still reluctantly began to acknowledge the usefulness of the non-traditional mathematical axiom, i.e., the axiom of choice, or its equivalence, the Zorn's Lemma.

The point of views expressed in this article could be quite personal until they get more acceptance from general audience. However, the readers will not be hurt by the author's viewpoints on mathematics expressed in this article but it will supply a fresh taste or sense to our main meal, mathematics.

2.1. Danger of Too Much Emphasis on Axiomatic Approach. One should note that those mathematicians, who introduced an appropriate set of axioms for the knowledge they already know, were also very good at computation, since they were trained as, called from now on, the "first" generation mathematicians, by their teaches most of whom were under the strong influence of computational approach of the 19th century. On the other hand, the "second" generation mathematicians, i.e., those who were mathematically trained by the first generation mathematicians, tended to start their mathematics not from heavy computational experiments but from a small amount of axioms, and they were more and more away from computational experiences. The lack of computational experiences is more significant in the "third" generation mathematicians, i.e., those trained by the second generation mathematicians.

Nevertheless, the strength of axiomatic approach had been fully demonstrated by the monumental works of a single mathematician, Alexander Grothendieck (1928 - 2014), on the Foundation of Algebraic Geometry, [**AG1**], FGA [**AG2**], EGA's [**AG3**] and [**AG4**], and SGA's [**AG5**] during 1950s up until the beginning of 1970s. His works in turn simply recall the works of a group of mathematicians, called Bourbaki [**B2**] and [**B1**], as elementary results. Anyone who is very much interested in the scope of axiomatic approach in mathematics explored during the 20th century is strongly encouraged to study and digest the above mentioned Grothendieck's works, of which the basic concepts and results are described in terms of categorical languages, called the **language of schemes**.

In particular, the tendency of the results obtained under a strong influence of axiomatic approach is too much conceptual, and hence becomes more and more **soft** and hence fragile, as time goes. In this case, a good-minded mathematician should worry about whether his and/or her results will still survive after, say, $1,000$ years from now. A prototypical example is Euclid's algorithm for computing the greatest common divisor of two numbers. His method is still known to us because it is about computation of certain mathematical objects in an efficient manner. Moreover, lack of computability only speeds up the isolation of Mathematics from other branches of Sciences. This is in no way acceptable to us.

For an elegant axiomatic treatment but still with a strong emphasis on computational point of views, of Algebraic Geometry, one may refer to Prof. George R. Kempf's two books [**GK1**] and [**GK2**].

3. A New Computational Tool: Computers

3.1. Personal Computers and High Performance Servers. Starting at around the beginning of the fourth quarter of the 20th century, computer industry has grown up with an unexpected rate of success; this nowadays allows us to compute very big numbers using a Personal Computer or even a Notebook Computer. Recently, inexpensive High Performance Cluster Servers are also available to us. As a tool for actual computation, computers are already playing a significant role and such role will surely change the mathematical world of the 21-st century. Here are some examples.

EXAMPLE A.1. *Here, $p(n)$ denotes, by definition, the number of ways of writing a given natural number n as a sum of others (called components) ignoring the order among the components. One starts with a very famous theorem:*

THEOREM A.1. *(Rademacher* [**GA**]*) Given a natural number* n*, one has*

$$p(n) = \frac{1}{\pi\sqrt{2}} \sum_{k=1}^{\infty} A_k(n)\sqrt{k}\frac{d}{dn}\frac{\sinh((\pi/k)\sqrt{(2/3)(n-\frac{1}{24})})}{\sqrt{n-\frac{1}{24}}}, \tag{A.1}$$

where

$$A_k(n) = \sum_{\substack{0 \le h < k \\ (h,k)=1}} \varpi_{h,k} \exp(-2\pi i n h/k),$$

where

$$\varpi_{h,k} = \exp(\pi i s(h,k)),$$

is a $24k$*th root of unity and where*

$$s(h,k) = \sum_{\mu=1}^{k-1} \left(\frac{\mu}{k} - \left[\frac{\mu}{k}\right] - \frac{1}{2}\right)\left(\frac{h\mu}{k} - \left[\frac{h\mu}{k}\right] - \frac{1}{2}\right)$$

is the so-called Dedekind sum.

It turns out that the value of $p(20,000,000)$*, when computed it using this theorem, requires more than* $4,900$ *number of decimal digits (Cf.* [**STS1**]*). Thus the final result is really a super-ultra big number, which is one point. Moreover, another point is that it only took about 17 Minutes with already-done pre-calculations of the so-called Dedekind sums, under the same hardware and software limitations as in the next example. These two points clearly show us the computing capability of current PC's.*

REMARK A.1. *One should note that computation of such a super-ultra number eventually raises a serious concern whether the computed result is absolutely correct. This is certainly a new kind of headaches to mathematicians.*

On the other hand, the author recently discovered an exact algebraic formula for a graded partition function $p_{\mathbf{M}}(n)$ *over a multiset* $\mathbf{M}$ *and an associated quasi-recursive formula for* $p_{\mathbf{M}}(n)$*, both of which* never *require approximation at all (Cf.* [**STS2**]*). One should note that it is neither quite* *possible to obtain an analytically exact formula for* $p_{\mathbf{M}}(n)$ *nor to obtain a recursive formula for it, something like either Rademacher's formula or the well-known Euler's recursive formula for* $p(n)$*, unless there is a significant symmetry among the elements of* **M**. *However, one version of the author's formulas computes all the values of* $p_{\mathbf{M}}(i)$ *for* $1 \leq i \leq n$ *with time efficiency of* $O(n^2)$ *for a general multiset* **M** *(Cf.* [**STS3**]*).*

3.2. Computer Algebra Systems. Of course, having just a very fast computing machine of which hardware is very reliable is not enough. One also needs a reliable software which guarantees several facilities. Among them, the most urgent requirements are, for instance, ability to handle numbers of any sizes and arbitrary precision in representing fractions, each of which absolute reliability is definitely required in any computation with computers.

EXAMPLE A.2. *The following Table shows the actual computation of the value of the classical partition function* $p(n)$*, for* $n = 200$*, using the Rademacher's formula in Eq (A.1). Here, by adding the first* 8 ***non-zero*** *terms of fractional real numbers, one gets a fractional real number,*

$$3972999029388.001346332,$$

which is very close to the true integer value $p(200) = 3972999029388$.

$$
\begin{aligned}
t(1) &= +3.9729989931858957578 72E + 12 \\
t(2) &= +36282.9784854411847056 \\
t(3) &= -87.58391287399 \\
t(4) &= +5.146534584 \\
t(5) &= +1.424217569 \\
t(6) &= +0.07102143214 \\
t(7) &= 0 \\
t(8) &= +0.04357655366 \\
t(9) &= +0.0256657539 \\
[t(10) &= +0.008603641042] \\
\hline
\text{Up to 9th Partial Sum} &= +3972999029388.001346332
\end{aligned}
$$

When it is about computation of an exact integer value in this way of approximation, there should not be any approximation-related errors in all of the intermediate steps. Do not expect that an error-free computation of such a kind is quite easy (Cf. [**STS1**]*), unless one uses very reliable Computer Algebra Systems.*

REMARK A.2. *This example, which is prepared by the author of this book, is especially interesting because it is a direct computational verification of the famous example computed by Hardy and Ramanujan (p. 284 of* [**HR**]*) and quoted by Andrews (p. 70 of* [**GA**]*). Indeed, the original calculation in Hardy's own handwriting occurs on the dust jacket of the 1969 edition of "A Mathematician's Apology",* [**H**]*. To our surprise, the calculation by Hardy and Ramanujan contains an error. The corrected calculation as in Example A.2 turns out to be much closer to* $p(200)$ *than the original incorrect version. Therefore, the third term,* -87.555*, in the original computation table for* $p(200)$ *in p. 284 of* [**HR**] *and later quoted in p. 70 of* [**GA**]*, needs to be replaced with* -87.584*.*

Any software package lacking these two capabilities should not be our choice for a tool for computational experiments. In addition to them, it is also required to have an ability to solve a given system of polynomial equations, i.e., non-linear equations over, say, the field of complex numbers. This means that it must have an ability to handle symbols as well as numbers. The theory for such computation had not fully understood until Bruno Buchberger (1942 -) discovered his famous algorithm for Groebner bases in 1965. Any reliable software package with at least these three facilities could be called a **Computer Algebra System**.

There are two types of Computer Algebra Systems, commercially available or not. Some of the commercially available ones are, MAPLE, MACSYMA, REDUCE, etc. Some of non-commercially available ones are, MACAULAY 2, CoCoA, etc. There should be more than these.

3.3. Behind Computational Experiments. One first notes that there are two kinds of computations one can do with computers, **exact computation** and **approximation**. Exact computation only needs addition, subtraction, multiplication, and allowed division depending on given conditions, on a given mixture of known symbols, unknown symbols, and numbers. In contrast to this, approximation allows us to apply non-exact division and limiting process, as well as addition, subtraction, multiplication and division on given numbers only. Thus, exact computation allows us to manipulate not only numbers but also **symbols**, whereas approximation only allows us to manipulate numbers. This tells us why exact computation is also called **Algebraic Analysis** or **Symbolic Analysis**, while approximation is called **Numerical Analysis**. Examples of Numeric Analysis are numerous. For instance, FEM (Finite Element Method) is just one of them. Also, see Example A.2 above.

These two approaches in computational methods do not hurt each other. In fact, they help each other. The only general trouble is that Algebraic Analysis does require some amount of elementary Algebra, mostly about polynomial rings in more than one unknowns, as well as some amount of Modern Algebraic Geometry, developed by Oscar Zariski [**OZ**] and J. P. Serre [**JS**], and a few others (For beginners, this amount of accounts on the subject may be good enough).

A proper sequence of computational approach is as follows:

(1) to apply appropriate exact methods in Algebraic Analysis until all available ones are exhausted, and then, as a final step,
(2) to apply appropriate approximation methods in Numeric Analysis.

The subject of Numeric Analysis has been well known to us after Sir Isaac Newton. However, Algebraic Analysis is very new to us. The most important theoretical background for recent revival of computational approach is the Buchberger's astonishing discovery of the algorithm for Groebner bases for polynomial ideals in more than one indeterminates. He figured out the method while he was working on his Ph.D. thesis, and it was first recognized by mathematicians oriented to computer science somewhat before or around 1975. I myself learned this remarkable algorithm in mid 1980s.

3.4. Danger of Too Much Emphasis on Computational Approach. If one puts too much emphasis on computational approach, he will soon find that he does not know why the method works and eventually he will get lost, because the arithmetic operations themselves never tell him what's going on, unless he knows *in advance* what's really going on. Unfortunately, it is obviously impossible most of the time, especially when one works on some mathematical phenomena not yet well understood to him. This means that one really needs a well-developed theory for exact computation or Algebraic Analysis.

There was no general theory for exact computation with computers until B. Buchberger discovered a very important algorithm. The mixture of his algorithm with the division algorithm produces very important algorithms for polynomial ideals in more than one variables. It is very urgent to understand this theory for hard but honest and diligent computation. The readers of this article should not underestimate the strength and usefulness of this theory.

4. Expected Trends of the 21st Century Mathematics

4.1. Computer Age. It is well known that there were two types of computers, analogue and digital, in the past. Nowadays, however, computers usually mean **digital computers** or their variants for better performance, since performance of digital computers (for instance, signal to noise ratio) has already proved to be a lot better than analogue types. It is also well known that the theoretical creation of digital computers was done by a famous mathematical logician, von Neumann, who invented Class Theory to answer, for instance, the question like whether the set of all sets is a set or not. After theoretically creating digital computers, he immediately developed a good deal of approximation theory for computing the values of non-polynomial functions with his theoretical machines.

Although Electric Engineering, Computer Engineering, and Computer Science are respectively needed to establish hardware, architecture, and operating system and software, the final combination of them is a system, called, computer, and the main reason for developing such a system is to compute or calculate something in any possible senses. The point we, mathematicians, should not miss is that **computation belongs to mathematics**. Thanks to computers, paperless computation of huge numbers is already available to those who know how to use some of the commercially or non-commercially available Computer Algebra Systems, so that the scope of Algebra and Algebraic Geometry in both pure and applied directions has already been expanded a lot.

4.2. Computers Are for Computation Related Human Activities. Since computation solely belongs to mathematics, it is mathematician's due obligation to examine and develop the needed mathematical algorithms in order to obtain from the machine the desired results with absolute correctness. In addition to this, our second obligation is to make it as efficient as possible. If it takes too much computation time, then such a computational approach will never buy other people's approval at all. To my regret, these two can't be easily accomplished by those who were not trained as mathematicians.

Unlike any other centuries, existence of so many mathematical objects have been proved during the 20th century, mostly thanks to active application of axiomatic methods influenced by Set Theory or Category Theory. These results are all inputs for further investigation that which of them are computationally constructible, applying the methods of Algebraic Analysis.

I should mention one dangerous aspect of the Zorn's Lemma or its equivalent axioms. If one used the Zorn's Lemma in some place of his proof of the existence of certain mathematical object, then his proof is computationally useless in the sense that it never really tells us how to construct such an object only using the four arithmetic operations and the limiting process. This is really a disappointing aspect of the Zorn's Lemma, but is, as long as **algorithmic construction** is concerned, still important enough to keep in mind in order for us not to use it in our proof as possible as we could.

Computers have already become a convenient and very capable tool for computations in any part of mathematics. We do not necessarily require to have an access to Supercomputers, Servers, or even Workstations for computational experiments. We only need to have Personal Computers for most computational experiments. Since Personal Computers are nowadays available to everyone and since they are already too fast and too capable of, most computational experiments could be done with them without spending too much time and money. Moreover, the hardware-related performance of these machines are nowadays very much reliable. The author strongly believes that such a very reliable computational environment, already available to everybody, sooner or later changes the whole atmosphere of mathematics of the 21-st century. For instance, the so-called **internet computing** will soon become very common.

One of the main ingredients of mathematics is perhaps its **exactness** in the process of both **proof** and **computation.** The point is that both of them require quite different types of exactness. This may be also the main reason why no other branches of Sciences even try to compete with mathematics, but only to rely on the results produced by mathematicians. Consequently mathematics has its own everlasting life. We should not give up any of these two powerful tools, proof and computation, to other branches of Sciences. On the other hand, we have been silently forced to ignore the significance of computational aspects for at least a century, mostly because of our inability in exactly hand-computing mixtures of big numbers and many parameters and also partly because of strong influences of axiomatic approach of the 20th century. Isn't it a time for us to revive computation related mathematical activities with a new tool, computers, to balance ourselves between them?

5. About Our InetCompu Service

We have recently started an E-mail based Internet Computing Service, called InetCompu.

If one visits

http://trinitas.mju.ac.kr/intro2InetCompu.html

the detailed information and several demonstrative services are provided:[1]

5.1. Our Concept and Terminology. If a person, for instance, using PC, PDA, or Mobile Phone, can perform the kinds of computations he wants on a remote Server, where he is not even required to have his own account, and is also allowed to get back the computed results immediately, then we call such a service, **InetCompu** (Internet Computing, Copyrighted) or **GridCompu** (Grid Computing, Copyrighted).

5.2. What We Have Done So Far. Several years ago, the author of this book figured out a practical method to actually realize the needed Software Interface (called General Purpose Grid Middleware Interface) for such an InetCompu Service, and he finished the development of the first primitive but working version of the Inferface at around August 16, 2001. Later on, the Interface for InetCompu was significantly upgraded by him to such a degree that it could handle the in-comming data more neatly and smartly. The final version currently in usage also passed a rigorous local fields test successfully.

As of August 28, 2001, we began to provide an InetCompu Service, i.e., whoever sends an E-mail to a particular E-mail address of our Server, for example, containing

(1) either an input data for an application program in our Server for a certain mathematical formula (for instance, input: n=1234567 to numbpart@trinitas.mju.ac.kr in plain text format

(2) or his own program, for instance, written in Reduce commands,

then he will soon receive the computed results from our Server.

[1]http://ch.catholic.or.kr/pundang/archive/4/intro2InetCompu.html

5.3. An Example of InetCompu. In particular,

http://trinitas.mju.ac.kr/intro2numbpart.html

explains how one can use our InetCompu Service to compute the value of $p(n)$ for a given input n, which utilizes Rademacher's formula for $p(n)$ in Theorem A.1.[2]

[2]http://ch.catholic.or.kr/pundang/archive/4/intro2numbpart.html

Bibliography

[AL] W.W. Adams and P. Loustaunau, *An Introduction to Gröbner Bases,* GSM Vol. 3, AMS, 1994.

[GA] G.E. Andrews, *The Theory of Partitions,* Reading, 2nd printing,1981, Addison-Wesley

[BW] T. Becker and V. Weispfenning, *Groebner Bases: A Computational Approach to Commutative Algebra*, Springer Verlag, 1993.

[B1] N. Bourbaki, *Algebra,* Part I and II (in English), Springer-Verlag, 1970, 1981.

[B2] N. Bourbaki, *Commutative Algebra* (in English), Springer-Verlag, 1985.

[CLO] D. Cox, J. Little, and D. O'Shea, *Ideals, Varieties, and Algorithms, An Introduction to Computational Algebraic Geometry and Commutative Algebra,* 2nd Ed., Springer-Verlag, 1997.

[ER] D. Eisenbud and L. Robbiano, *Computational Algebraic Geometry and Commutative Algebra,* Symposia Mathematica Vol. XXXIV, Istituto Nazionale di Alta Mathematica Francesco Severi, Cambridge Univ. Press, 1993.

[DE] D. Eisenbud, *Commutative Algebra with a View Toward Algebraic Geometry,* Springer-Verlag, 1995.

[E] Euclid, *The Elements,* Vol. 1 - 3 (English translation), 2nd Ed., Dover, 1956.

[AG1] A. Grothendieck, *Algébre Homologique,* Secrétariat mathématique, 11 rue Pierre Curie, PARIS 5e, 1958.

[AG2] A. Grothendieck, *Fondements de la Géométrie Algébrique* (Extracted from Bourbaki Seminar 1957-1962), Secrétariat mathématique, 11 rue Pierre Curie, PARIS 5e, 1962.

[AG3] A. Grothendieck and J.A. Dieudonné, *Eléments de Géométrie Algébrique,* Vol. I-IV, IHES, 1960-1967.

[AG4] A. Grothendieck and J.A. Dieudonné, *Eléments de Géométrie Algébrique I,* Springer-Verlag, 1970.

[AG5] A. Grothendieck, *Séminaire de Géométrie Algébrique,* Vol. 1-7, IHES, 1960-1962.

[HR] G. H. Hardy and S. Ramanujan, *Asymptotic Formulae in Combinatory Analysis*, Proc. London Math. Soc. (2), Vol. 17 (1917), pp. 75-115. (Reprinted in

Collected Papers of S. Ramanujan, pp. 276-309, Cambridge University Press, London, 1927.)

[H] G.H. Hardy, *A Mathematician's Apology*, Cambridge University. Press, 1969.

[GK1] G. Kempf, *Algebraic Varieties*, London Math. Soc. LNS 172, Cambridge Univ. Press, 1993.

[GK2] G. Kempf, *Algebraic Structures,* Friedrich Vieweg & Sohn, 1995.

[DM] D. Mumford, *Algebraic Geometry I, Complex Projective Varieties,* Springer-Verlag, 1976.

[JS] J. P. Serre, *Faisceaux Algébriques Cohérents*, Annals of Mathematics, Vol 61, No. 2, March, 1955.

[STS1] S.T. Soh, *A Reliable Implementation of the Formula of Rademacher for the Partition Function $p(n)$*, preprint

[STS2] S.T. Soh, *A Diophantine Construction of an Exact Algebraic Formula for Graded Partition Functions*, J. Korean Math. Soc., Vol 36 (1999), No. 2, pp. 267-298

[STS3] S.T. Soh, *Efficient (Quasi-) Recursive Formulas for Graded Partition Functions,* preprint

[ES] E.H. Spanier, *Algebraic Topology,* Springer-Verlag, 1966.

[BS] B. Sturmfels, *Algorithms in Invariant Theory,* Texts and Monographs in Symbolic Computation, Springer-Verlag, 1993.

[BW] B.L. van der Waerden, *Algebra,* Vol. 1 and 2, Frederick Ungar Pub., 1970.

[WR] A.N. Whitehead and B. Russell, *Principia Mathematica*, Cambridge Univ. Press, 1910.

[OZ] O. Zariski, *Collected Papers,* Vol. 1 - 4., MIT Press, 1972.

[ZP] O. Zariski and P. Samuel, *Commutative Algebra,* Vol.1 and 2, Springer-Verlag, 1958.

APPENDIX B

Suggestions for Further Reading and References

M. Artin, *Algebra*, Prentice Hall, 1991.

G. Birkhoff and S. MacLane, *A Survey of Modern Algebra*, 4th Edition, Macmillan, New Nork, 1997.

D. Cox, J. Little, and D. O'Shea, *Ideals, Varieties, and Algorithms*, 2nd Edition, UTM, Springer-Verlag, 1997.

J. Fraleigh, *A First Course in Abstract Algebra*, 7th Edition, Addison-Wesley, 2000.

A.C. Hearn et al, *REDUCE User's Contributed Package Manual*, Version 3.8, Konrad-Zuse-Zentrum, Berlin, 2004.

T. Hungerford, *Algebra*, Graduate Text in Mathematics, Springer-Verlag, 1974.

G. Kempf, *Algebraic Varieties*, London Math. Soc. LNS 172, Cambridge Univ. Press, 1993.

G. Kempf, *Algebraic Structures,* Friedrich Vieweg & Sohn, 1995.

O.T. Kim and S.A. Park, *Modern Algebra* (Korean), 3rd Edition, Cheongmungak, 1993.

M. Kofler, *Maple: An Introduction and Reference*, Addison-Wesley, 1997.

S. MacLane and G. Birkhoff, *Algebra*, 2nd Edition, Macmillan, New York, 1979.

S.T. Soh, *Linear Algebra*, Kyowoo, Seoul, 2005, 2019 (2nd Edition).

S.T. Soh, *Computational Algebra*, Kyowoo, Seoul, 2006.

S.T. Soh, *Algebraic Analysis I*, Kyowoo, Seoul, 2006.

B.L. van der Waerden, *Modern Algebra*, Ungar, New York, 1970.

O. Zariski and P. Samuel, *Commutative Algebra,* Vol.1 and 2, Springer-Verlag, 1958.

http://trinitas.mju.ac.kr/intro2InetCompu.html

http://ch.catholic.or.kr/pundang/archive/4/intro2InetCompu.html

http://www-groups.dcs.st-and.ac.uk/~history/

Index